Historical & Cultural Astronomy

Historical & Cultural Astronomy

More information about this series at http://www.springer.com/series/15156

Ileana Chinnici
Editor

Merz Telescopes

A Global Heritage Worth Preserving

Springer

Editor
Ileana Chinnici
INAF-Osservatorio Astronomico di Palermo
Palermo
Italy

ISSN 2509-310X ISSN 2509-3118 (electronic)
Historical & Cultural Astronomy
ISBN 978-3-319-41485-0 ISBN 978-3-319-41486-7 (eBook)
DOI 10.1007/978-3-319-41486-7

Library of Congress Control Number: 2017930282

Printed on acid-free paper

This Springer imprint is published by Springer Nature
The registered company is Springer International Publishing AG
The registered company address is: Gewerbestrasse 11, 6330 Cham, Switzerland

Preface

It is undeniable that Merz telescopes played an important role in the development of astronomy and astrophysics in the nineteenth century. The reliability of these instruments, whose excellent optical qualities were universally recognized, is proven by their widespread use: in the second half of the nineteenth century, there were no observatories devoid of a Merz instrument, be it a large or a medium-size one, depending on the scientific program and/or the budget of that establishments. Merz telescopes were installed in almost all the most important observatories, in Europe, but also in India, the Philippines, Japan, Ecuador, and many other countries: their contribution to the diffusion of the practice of astronomy is unquestionable. Moreover, these instruments were active for incredibly long periods of time, having been used for over a century in many cases, sometimes even recycled, modernized, or combined with other telescopes, especially in regard to their optical parts.

This happened in Italy, for example, where Merz telescopes or lenses were used until the 1970s. Actually, Italy was one of the main commissioners of the Merz Company, due to the fact that, after the political unity reached in 1861, and the annexation of Rome in 1870, the Italian government had to sustain most of the astronomical observatories in existence in the territories of the previous Italian States. Many of these observatories had acquired or were already equipped with Merz telescopes. In the second half of the nineteenth century, early spectroscopic studies on stellar and solar physics were carried out in some Italian observatories, with Merz instruments being of primary use for this kind of research. In the twentieth century, as the financial resources allocated by the Kingdom of Italy for the renewal of astronomical equipment were scarce, Merz instruments still in use were often modernized, in order to continue to exploit their excellent optical performances. Sometimes, the disassembled lenses were mounted in other telescopes or given on loan for scientific purposes: many cases of "cannibalization" have been recorded but many others have probably been forgotten and, consequently, many pieces have been lost.

In order to preserve this important and unique heritage, actions have been taken in the past 20 years, by some Universities and astronomical institutions, for inventorying, restorying and displaying these instruments in museums. Little by little, in Italian astronomical observatories, an increasing awareness of the importance of preserving this heritage has developed among astronomers, and today, the risk of destroying or throwing away old instruments, books and papers—as often happened in the past—has been almost completely eliminated.

This book intends to reinforce the consciousness of the scientific community about the value of this heritage and stimulate initiatives aimed at preserving Merz telescopes in other observatories and countries. It offers a perspective about the results of these actions in Italy and the research work requested for this purpose. The initial chapters provide a general view of the Merz Company and the building of large refractors in the nineteenth century, while the subsequent chapters deal with specific aspects, collections and instruments and the studies carried out with them, up to recent times. The contributors are experts in the field and often curators of the collections herein described.

As a final remark, it is important to stress that the importance of this heritage is sometimes underestimated. Historical instruments are often regarded as a mere resource for popularizing astronomy. This is certainly a part of their value, since they represent a powerful tool for inducing people to engage with science, by means of storytelling, and a visualization of how science has advanced thanks to new ideas and technologies. Nonetheless, this is a restrictive perspective. Scientific material heritage, above all, consists of cultural goods to be preserved, and is a tool for historical research. Around every old instrument, in fact, there is an intertwined context of ideas, persons, situations, and institutions. Knowing this background is crucial to fully understanding the evolution of science, and adjusting the mistakes that have sometimes occurred because of an overly facile approach to history; correct contextualization of these objects enables everyone, scholars and public, to gain the right perspective about a theory, a discovery, or a way of engaging with science. Retaining memory also means preserving identity: indeed, looking at the past and studying scientific heritage is itself most assuredly a contribution to science.

Palermo, Italy Ileana Chinnici

Contents

Abbreviations

AAVSO	American Association of Variable Star Observers
APUG	Archivio Pontificia Università Gregoriana
ASO	Attività Scientifica, Osservazioni
ASS	Attività Scientifica, Strumenti
C	Correspondence
FA	Fondo degli Astronomi
FS	Fondo Secchi
GBD	Giovanni Battista Donati
HAOA	Historical Archives of Arcetri Astrophysical Observatory
HAOC	Historical Archives of the Astronomical Observatory of Capodimonte
HAOPd	Historical Archives of Padua Observatory
INAF	National Institute for Astrophysics
MPI	Ministero della Pubblica Istruzione
Ms	Manuscript
MuSA	Museum of the Astronomical Instruments, Capodimonte Observatory
OACt	Catania Astrophysical Observatory
OAPa	Palermo Astronomical Observatory
OAR	Rome Astronomical Observatory
OAT	Turin Astronomical Observatory
SAN	State Archives of Naples
SCA	State Central Archives
SLF	Sede, Lavori di Fabbrica
StAM	Stadtmuseum Munich

Big Is Beautiful: A Few Considerations About the Making of the Large 19th Century Refractors

Paolo Brenni

The race to produce very large refractors was one of the characteristics of 19th century instrumental astronomy. Much has been written about the history and evolution of these telescopes, and several lists of them have been compiled.[1] In 1823 Joseph von Fraunhofer completed the largest refractor of his time: the equatorial telescope for the Dorpat observatory. This instrument, with its special mounting (German mounting), a large and excellent objective, and a clockwork driving mechanism, proved to be a great success. In the following decades, G. Merz who was Fraunhofer's successor, constructed many instruments of this type. Until the beginning of the 20th century, large German mounted telescopes were, together with meridian circles, the most important instruments of astronomical observatories. German, British, French and American engineers and opticians competed in order to produce the largest and more powerful equatorials. In this article I describe some of the problems related with the construction of these instruments and I describe the characteristics of some of the most important ones.

[1]About the history of large refractors, see, in particular: Van Helden (1985), Lequeux (2009). About the use and evolution of telescopes in general, see: Repsold (1914), Danjon, Couder (1979), King (1979), Bell (1981) Riekher, Beck (2012). Detailed technical descriptions of large 19th-century telescopes can be found in Ambronn (1899). See also: www.astrosurf.com/re/building_large_telescopes_refractors.pdf (accessed 25 October 2016).

P. Brenni (✉)
CNR, Fondazione Scienza E Tecnica, Florence, Italy
e-mail: pbrenni@museogalileo.it

I. Chinnici (ed.), *Merz Telescopes*, Historical & Cultural Astronomy,
DOI 10.1007/978-3-319-41486-7_1

1 The Prototype: Dorpat Refractor

It is generally accepted that the beginning of the saga of the great 19th century refractors coincides with the realization of the one made by Joseph von Fraunhofer (1787–1826) for the Observatory of Dorpat (today Tartu, Estonia). In 1824, Friedrich Georg Wilhelm von Struve (1793–1864), the director of the observatory, while admiring the assembled telescope, wrote:

> I was amazed standing in front of the marvellous masterpiece and I was unable to decide what I found most admirable: the beauty of the forms of the whole piece and its completion to the smallest details, or the effectiveness of the installation and the useful device to facilitate movement, or the unsurpassed optical quality and precision of the images. [...] After the proper correction of the counterweights, the assembled instrument is perfectly balanced in every position. With a finger, it is possible to move it along the axis parallel to the Equator and, with an even smaller force, along the axis of the Earth [...]. So the huge tube can be moved along the horizon very rapidly and securely in every desired position (Struve 1826).[2]

Moreover, the optics of the instrument fully satisfied Struve, who added: *So I absolutely consider our achromatic lens the most perfect masterpiece of optics that has ever existed* (Struve 1826).

This telescope certainly was the mechanical and optical marvel of the time, and today is rightly considered the father of all the large 19th century refractors (Fig. 1). The instrument has several peculiar and important characteristics:

(a) An equatorial mounting which became known as a German equatorial mounting (or simply a German mounting). It consisted essentially of a T-shape structure, whose lower bar, aligned with the celestial pole, is called the right ascension axis and the upper transversal bar is the declination axis. The telescope was placed on one end of the declination axis. This mounting also allowed observations in the direction of the celestial pole.
(b) Two large divided circles with verniers for reading scales.
(c) A centrifugal driving mechanism that allows the device to follow a celestial body automatically, by compensating for the apparent movement of the sky.
(d) A perfectly balanced wooden tube, having two parallel bars with counterweights which reduce its flexion.

---

[2]*Stauned stand ich nun vor dem herrlichem Kunstwerke, unterschieden, was mehr zu bewundern sey, die Schönheit der Formen des Ganzen und die Vollendung derselben bis in kleinste Detail, oder die Zweckmäfsigkeit der Aufstellung und der so sinnreiche Mechanismus zur Bewegung, oder die unvergleichliche optische Kraft und Präzison. [...] Das so zusammengesetzte Instrument ist nach gehöriger Berichtigung der Gegengewichte in jeder Lage in völligem Gewicht.Mit einem Finger kann man es um die im Aequator liegende Achse drehen, mit einer noch weit geringeren Kraft um die Welt-Achse, [...] So lässt sich dies gewaltige Rohrmit wei grösserer Schnelligkeit und Sicherheit in jede beliebige Lage [...]. Somit möchte ich unbedingt unserer Achromat für das vollkommenste Kunstwerk der Optik, das bisher existiert hat, halten.*

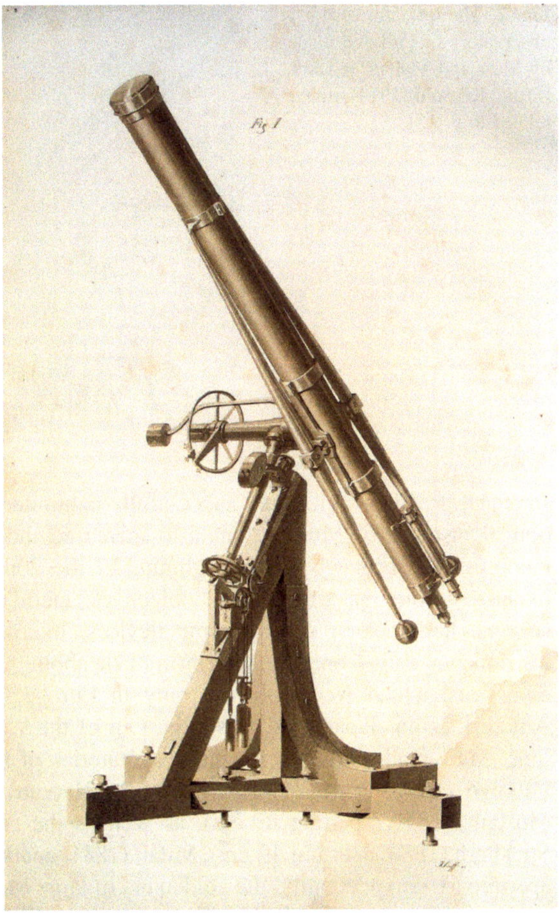

Fig. 1 The 24 cm refractor made by Fraunhofer for the Dorpat (Tartu) astronomical observatory (From: Struve 1825)

(e) A large (24 cm of aperture) and perfectly executed achromatic objective with a series of eyepieces, including a micrometric eyepiece. This was the largest objective-lens of that time.

Certainly, a number of these elements were not new and had been included in other, older, instruments, but, for the first time, they were combined together in a remarkably large, very efficient and powerful refractor. The instrument proved to be successful and, soon, other similar and larger instruments were constructed.

2 Merz Refractors

Fraunhofer started to build another refractor, similar to the one at Dorpat, for the Berlin Royal Observator but, in 1826, his premature death, at the age of 39 years, put a sudden stop to his brilliant career. However, Fraunhofer's legacy was not lost. First, Georg Merz (1793–1867) and Joseph Mahler (1795–1845), and then Georg's

Fig. 2 The refractor of the observatory of Pulkovo made by Merz and Mahler in 1839. [From: Repsold (1914), image related to p. 15]

sons and grandsons, not only successfully continued the production at the famous optical institute of Munich, but also increased the importance and enlarged the market of the firm, up to the beginning of the 20th century.[3] Merz's production included refractors, astronomical objectives, meridian and vertical circles, comet seekers, heliometers, spectroscopic devices, microscopes, binoculars, surveying instruments, and so on. The Merz firm made about 90 large refractors and 100 large objectives, which were sold not only in Europe, but also in North and South America, India, Japan, and Australia. Out of the very large refractors built at the time, Merz made the ones for the observatories of Berlin (1829, aperture 24 cm), Pulkovo (1839, aperture 38 cm) (Fig. 2), Harvard (1847, aperture 38 cm), and Moscow (1859, aperture 49 cm), as well as the objectives for the refractors of Strasburg (1880, aperture 49 cm), Milan (1881, aperture 49 cm), and Manila (1893, aperture 51 cm). Certainly, the golden era of large Merz refractors covers the period between the end of the 1830s and the early 1880s. After that time, the firm built a few more instruments of this kind or supplied large objectives to others.

With the increasing focal length and aperture of the large equatorials, the structure of the instruments became more complex and required technical solutions which an optician and instrument-maker could no longer master. This is one of the reasons why, in the second half of the 19th century, the Merz firm, which also refused to hire people outside Bavaria, lost its leading position in the construction of large telescopes. Indeed, it continued to supply large and excellent objectives, but for the realization of the mechanical parts of refractors, it was necessary to turn to specialized engineering firms. Late 19th century refractors were heavy machines and, at the same time, delicate instruments. The massive mountings, the big counterweights, the robust axles and their bearings, the long metallic tubes, all were typical products of heavy mechanical engineering. For example, the steel polar axe of the Yerkes Observatory's main telescope was more than 4 m long and weighed

[3]For detailed history of the Merz firm and its instruments, see Kost (2015).

Fig. 3 Hoisting the polar axis of the Yerkes observatory 40-inch refractor telescope. [University of Chicago photographic archive, [apf6-00033], special collections research center, University of Chicago Library]

3.5 tons, while the declination axe was about 335 cm long and weighed 1.5 tons (Fig. 3). Nevertheless, the large divided circles and their reading microscopes, the sophisticated driving mechanism that compensated for the movement of the Earth and kept the observed object in the field of vision of the telescope so that good astronomical photographic plates could be taken, were elements requiring the precision work typical of the high quality instrument industry. Moreover, good and very large achromatic objectives could be ground and polished only by a few very skilled and very specialized opticians who had to order the best glass blanks (often weighing several hundred kilograms) from the only three or four glassmaking companies capable of supplying them. Therefore, while the Merz firm had, for a good number of years, been able to produce the optical glass and fashion the large lenses, as well as the mechanical parts of the telescope, in the second half of the 19th century, very large refractors were often made thanks to the collaboration of different actors (engineers and instrument-makers, opticians and glassmakers). Among them, a few names have particularly to be recalled.

3 Collaborators and Rivals

In Germany, the famous firm founded by Johan Georg Repsold (1770–1830) in 1799, and continued by his descendants up to 1919, was responsible for the construction of the mountings for the large refractors of Strasburg (1880), Pulkovo

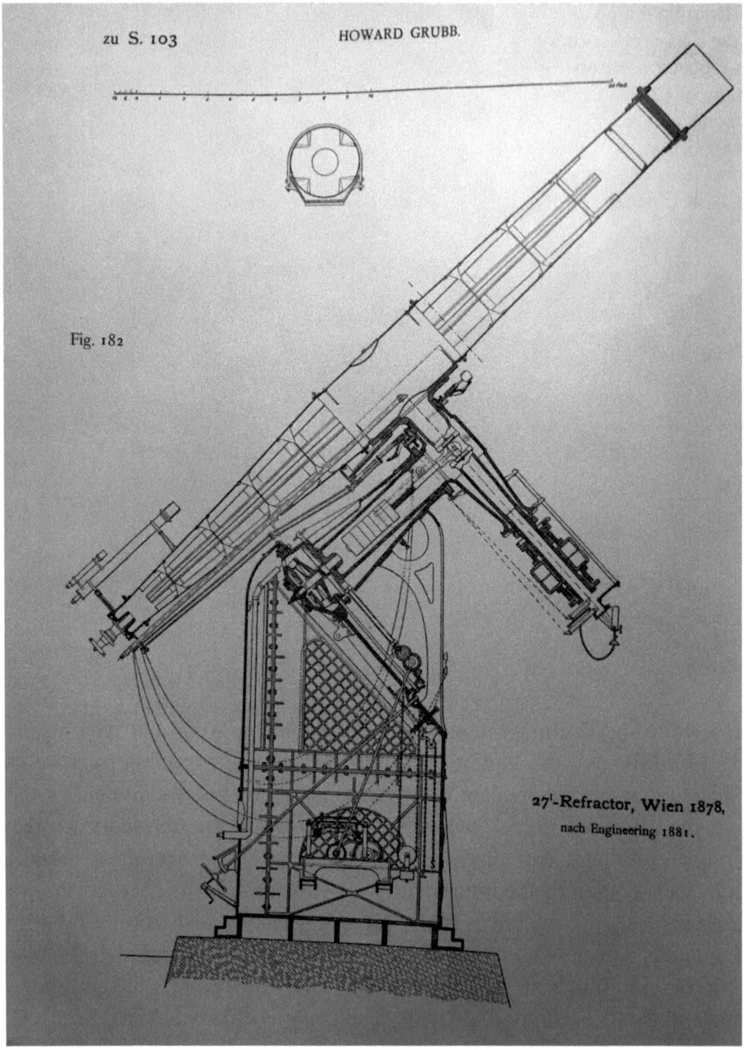

Fig. 4 A technical scheme of the refractor of the observatory of Vienna, made by Grubb in 1880. [From: Repsold (1914), image related to p. 103]

(1884) and Potsdam (1899). The firm of the Irish engineer Thomas Grubb (1800–1878) and his son Howard (1844–1931) made the equatorials for Cambridge (1861, aperture 63 cm), Vienna (1880, aperture 69 cm) (Fig. 4), Daramona (1881, aperture 61 cm) in Ireland, Greenwich (1896, photographic objective of 66 cm) and Cape Town (1901, photographic objective of 61 cm). In 1869, the renowned English firm *Thomas Cooke and Son* made a refractor with an aperture of 64 cm for the wealthy amateur astronomer Robert S. Newall (1812–1889): it was, for a short period, the largest of its time (Fig. 5).

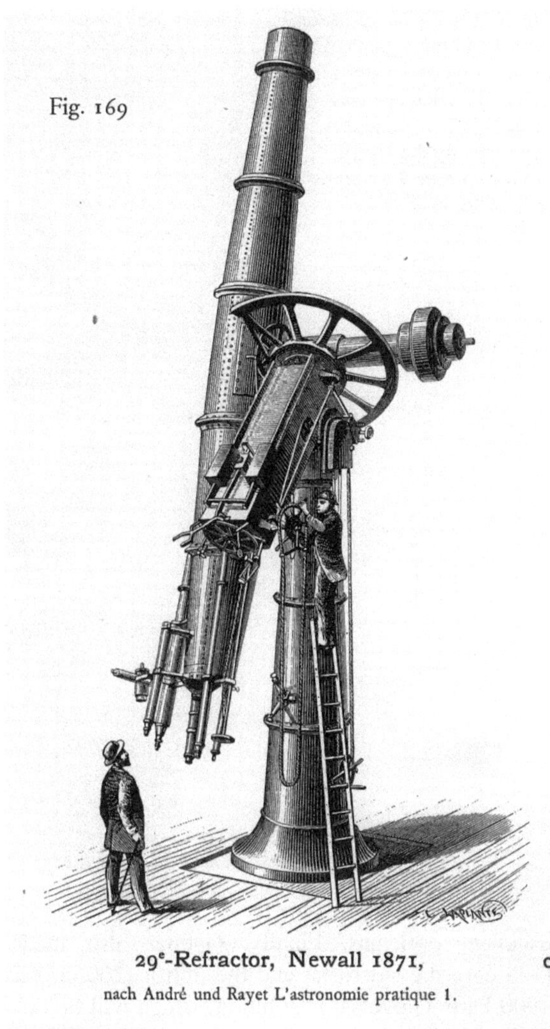

Fig. 5 The Newall refractor made by Cooke in 1869 for the Newall private observatory. In 1890, the refractor was offered to the University of Cambridge and, in the late 1950s, was donated to the national observatory of Athens. [From: Repsold (1914), image related to p. 99]

In France, the most famous builder of large equatorial telescopes was Paul Gautier (1842–1909), who constructed mountings for the refractors of the Nice Observatory (1887, aperture 76 cm), for the double refractor of the Meudon Observatory (1869, aperture 83-cm visual and 62-cm photographic objectives) (Fig. 6), for the Zo-Se Observatory (China, 1897, 40 cm visual and photographic), and for the Athens Observatory (1899, aperture 40 cm), as well as for the peculiar Loewy's coudés telescopes installed in France (at the Paris, Nice, Lyon and Besançon observatories) in Vienna and in Algiers. Most of the largest objectives mounted on Gautier's refractors were made by the French brothers Paul Pierre (1848–1905) and Prosper Mathieu Henry (1849–1903), who were astronomers and

Fig. 6 The double refractor (visual and photographic) of the Meudon observatory in 1896. The telescope was made by Gautier and the objectives by the Henry brothers [From: *La Nat*ure, 1 sem. 1896, p. 361]

excellent opticians. Finally, Gautier also made the photographic objective (125 cm), the siderostat and the mirror (200 cm) for the gigantic telescope of the 1900 Paris Universal Exhibition, which will be discussed below.

In the United States, Alvan Clark (1804–1887) and his sons were also famous for various very large refractors and objectives. In 1862, they made the 47-cm equatorial for the Dearborn Observatory (Illinois), and in 1863, the one for the Cambridge (Massachusetts) Observatory (aperture 63 cm) and the one for the Vienna Observatory (aperture 69 cm in 1880). In 1891, they also made the refractors for the Cape Town and Oxford Observatories (both with 61 cm photographic and 46-cm visual objectives), and in 1896, the 61-cm telescope for the Flagstaff Observatory. Finally, the same firm also designed the large objectives for the refractors of the U. S. Naval Observatory in Washington (1877, aperture 66 cm), the Lick Observatory in California (1888, aperture 91 cm) and the Yerkes Observatory in Illinois (1897, aperture 102 cm, Fig. 7). The mountings for these

Fig. 7 The 102-cm Yerkes refractor in 1997. (University of Chicago Photographic Archive, [apf6-00312], special collections research center, University of Chicago Library)

three instruments were all made by the American engineering firm *Warner and Swasey*.[4]

4 Glassmakers and Optical Glassmaking

If only a few European and American firms were capable of constructing very large refractors and fashioning their lenses, there were even fewer glassmakers who could provide the high quality optical glass that they needed. The Merz firm, following the path opened in Benediktbeuern by Fraunhofer and the Swiss glassmaker Pierre Louis Guinand (1748–1824) at the beginning of the 19th century, produced its own glass in the same glassworks until 1883. But up to about 1900, the other most important optical glassmakers also had a more or less direct connection with Guinand. In fact, in the late 1820s, his son Henri (1771–1852) disclosed the art of optical glassmaking to George Bontemps (1801–1882), who sold the secret technique to the glass factory of the Chance Brothers in Birmingham, ultimately joining

[4]A detailed chronological list of the largest refractors of the 19th and early 20th century can be found in Lequeux (2009). A similar list can also be found in Duval (1896) at: https://en.wikipedia.org/wiki/List_of_largest_optical_refracting_telescopes (accessed 25 October 2016) and also http://central.gutenberg.org/article/whebn0000213473/list%20of%20largest%20optical%20refracting%20telescopes (accessed 25 October 2016).

that company. On the other hand, Charles Feil, Henri Guinand's grandson, continued the family tradition in France. His firm passed to Edouard Mantois (1848–1900) in 1887, eventually being renamed Parra-Mantois around 1900. We can affirm that, except for those built by Merz, almost all the optical glass blanks used for the large objectives, all around the world, were produced in Birmingham by the Chance Brothers, or in France by Feil and Mantois and their successors. It is worth noting that the famous Schott glass company was originally founded in 1884 to serve just as a glass research laboratory and, with very few exceptions, the Jena glass began to be used for large astronomical instruments only from the very late 19th century. In fact, before 1900, Schott only provided the blanks for the optics of the telescope at the Archenhold Observatory in Berlin (1896, aperture 68 cm) and for the double refractor at Potsdam (1899, 80-cm photographic and 50-cm visual objective). The lenses of those two instruments were made by the Steinheil firm.

However, in spite of the great progress made since the time of Fraunhofer and Guinand, the production of large optical glass blanks suitable for being transformed into large lenses remained a very difficult task, even at the end of the 19th century. The above-mentioned glassmaker Mantois wrote an interesting article describing the difficulties in obtaining the suitable glass (Clark and Mantois 1894). For example, the finished flint glass element of the achromatic doublet of the Yerkes telescope weighed about 160 kg, but the blank from which it was made weighed more than 400 kg. Working such a mass of glass was extremely delicate and painful. First, the crucible, heated in a special oven, had to be filled with a mixture of sand, potash, boric acid, lime, red lead, etc. The operation had to be carried out slowly (15–18 hours) and often had to be repeated a few times, because the melted mixture produced bubbling and risked overflowing from the crucible into the oven (Fig. 8). The mass was heated for 20–30 hours at a very high temperature. This operation was dangerous, because the crucible could crack and the refractory bricks could melt. After that, it was necessary to start the "*brassage*": the melted glass had to be stirred to amalgamate all the components in order to obtain a homogeneous fluid. That was done with a tool, made of a clay cylinder with a long iron bar and a wooden handle. The operation was exceedingly painful and very long (10–15 hours). The heat near the crucible was so unbearable that the team of four workers moving the tool, under the supervision of the chief glass-maker, had to be rotated every 5 minutes, even if they were wearing special gloves and aprons of asbestos. From time to time, the molten glass produced bubbles, and the crucible had to be closed and reheated. Then, the temperature was slowly lowered and the *brassage* continued, but the operation became very difficult, due to the fact that the molten glass became more and more viscous. When the operation was judged complete, the mass was cooled. To avoid dangerous internal stresses in the glass, the cooling had to be done very slowly in a sealed oven and could last from four to six weeks. Finally, the oven was opened and the pieces of glass were taken out of the broken crucible. In spite of all the care and the long *brassage*, an optical exam often showed that only about 50% of the glass was good and homogeneous enough to be used in the making of a large lens. The best blanks were selected and placed in special moulds. These were put in a low temperature oven so that the glass could be

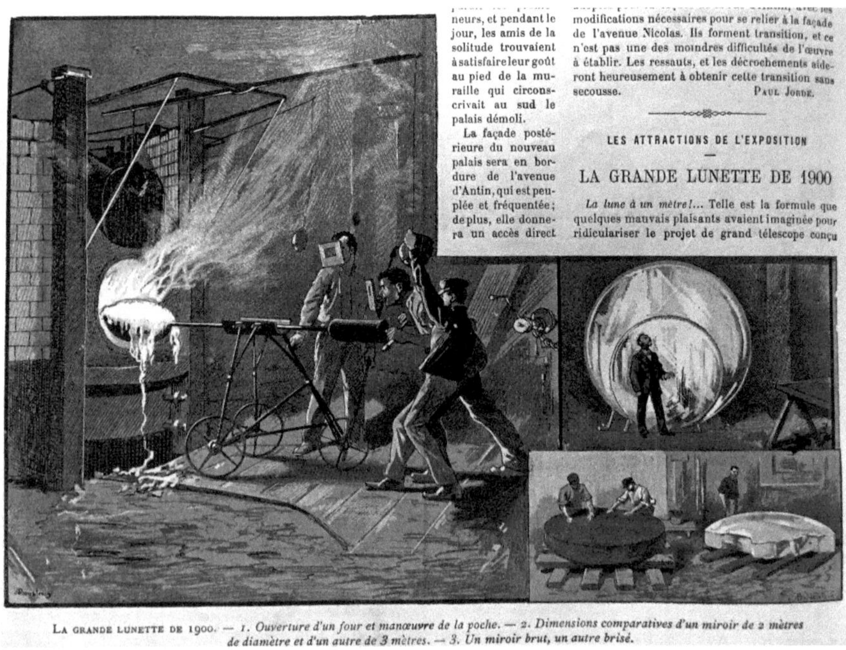

Fig. 8 The making of optical glass for the 1900 exhibition telescope. [From: *L'Exposition de Paris (1900)*, Paris, Mongredien et Cie, 1900, vol. 1, p. 110]

softened (but not melted again!) in order to take on the form of the moulds. Again, after a few weeks of cooling, a polygonal glass disk could be removed from the mould. It was not uncommon, during the change in temperature, for the glass, to break or crack (that happened more than once during the preparation of the blanks for the Yerkes telescope, for example). The glass disk was examined (after the faces had been sawed and polished, Fig. 9), and other striae or bubbles often became visible. One could try to eliminate them, by sawing away the inhomogeneous portions of the disk. The blank had to be replaced in its mould and put in the oven again, and the operations were repeated. The last moulding had to confer upon the blank the approximate shape of the desired lens: the mould weighed 500 kg and its preparation took two months of work. After 10 days, the softened blank had taken the shape of the lenticular mould and was again slowly cooled down for more than three weeks. Finally, the blank was examined a last time and proved to be ready for polishing. The entire operation lasted about 17 months!

Grinding and polishing a large lens was also a very difficult task. The ideal curvatures of flint and crown lenses in a doublet should have been corrected for chromatic and spherical aberrations over the entire spectrum (not to mention astigmatism and coma). Practically, the chromatic aberration of an objective used for visual observations was corrected for the region of the spectrum to which the eye is more sensitive, while for the objectives intended for photography, the corrections were done for the violet part of the spectrum. Spherical aberration was

Fig. 9 Sawing a blank of optical glass at the end of the 19th century, postcard of Parra-Mantois et Cie (courtesy of Archives de Saint-Gobain)

Fig. 10 The large grinding and polishing machine used by Gautier for making the lenses of 1900 exhibition telescope. [From: *L'Exposition de Paris (1900)*, Paris, Mongredien et Cie, 1900, vol. 2, p. 66]

corrected for several wavelengths in choosing the best curvature for each lens. However, the simultaneous minimization of various aberrations required a delicate compromise, which was the result of a combination of mathematical calculations and practical knowhow. For grinding and polishing large telescope lenses, special steam-driven (or later electric) machines were used (Fig. 10). These were a kind of large, vertical-axle optician lathe, with a turntable supporting the blank. The tool acting on the surface of the glass was moved by hand or by a system of articulated, eccentric arms, acted on by a motor. The quality of a lens could be checked by observing either a real or artificial star, but in 1858, Léon Foucault (1819–1868) introduced the famous knife-edge test, which produces an image showing the characteristics of the surface and its imperfections.

Once the desired curvature had been obtained, it was necessary to correct the smallest defects on the surface. Only the best opticians, such as the afore-mentioned Merz, Clark, Grubb, the Henry brothers and a few others, were able to do the final retouches. Thanks to an incomparable practical knowhow acquired after years of experience, they corrected the local imperfections by hand, after a long and accurate examination of the surface. Sometimes, they would rub the area to be corrected several times with a finger. During the final phases of polishing, even the slightest friction could produce a thermal expansion of the glass and, before checking the lens again, it was necessary to wait until it reached a perfect thermal equilibrium. Consequently, it is not surprising that very large lenses took years to be delivered and were very expensive. For example, the objective of the Yerkes telescope cost US$66,000 and the one at Lick, US$50,000, while the mountings for these telescopes, respectively, cost US$55,000 and US$44,000 (Hussey 1897).[5] To these figures, one had to add the expenses necessary to build a dome and its ancillary equipment. In fact, the buildings housing these instruments were also complex machines. The refractor had to be installed at the top of a concrete pier, separated from the rest of the building in order to avoid vibrations. The moveable metallic domes represented an engineering tour de force. In spite of its diameter and weight, it had to turn easily and smoothly. At the Nice Observatory, the dome, made by the famous Gustave Eiffel (1832–1923), was floated on an annular vessel filled with water, while in Yerkes, it was supported by 36 wheels with roller bearings. Special ladders and elevator chairs allowed astronomers to reach the eyepiece of the instrument, but the entire floor of the building could often be lifted and lowered with a series of mechanical contrivances so that the astronomer could easily look into the eyepiece independently of the orientation of the telescope.

5 Some Unsuccessful Cases

Not all of the large 19th century refractors were successful. Some of them failed because of ill-conceived construction or the poor quality of their objectives, while others were underexploited because of the lack of a well-defined research project. A few examples illustrate these facts.

In the early 1850s, the Italian topographer and instrument-maker Ignazio Porro (1801–1875), who had his workshop in Paris, opened his “astronomical park” there. In it, he installed several instruments, among which there was an unusually large altazimuth refractor with a 52-cm aperture lens (Fig. 11). Porro not only claimed

[5]It is very difficult to calculate the value of 1900 dollars into today’s dollars. Many factors, such as the change in cost of labour, have to be considered. However, depending on the chosen criteria, the relative value of US$50,000 in 1900 ranges from one to several millions of dollars of today.

Fig. 11 The 52 cm refractor with azimuthal mounting installed by Ignazio Porro in his Parc Astronomique in Paris. [From: Porro (1857), p. 8]

that it was the best and largest objective of its time, but also that he had invented an unsurpassable method for making perfect lenses. The result of an examination, conducted by some members of the Académie des Sciences, however, was far from positive. Finally, when Porro left Paris in 1859, the refractor was dismantled (Brenni Ms, Porro 1857).

Around 1850, the Reverend John Craig (1805–1877) spent a fortune for installing another "monster refractor" near London. It had a 61-cm achromatic doublet (the largest of its time) and a tube of 30 meters. The mounting of the instrument was very unusual, and certainly very unpractical. The tube was fixed to the hoop at one end of a huge chain, slung from the side of a tower. The eyepiece end of the tube rested on a movable wooden dolly on the azimuth arm. But the instrument turned out to be far too difficult to be orientated and used, and it did not fulfil the expectation of its owner. Eventually, the telescope was dismantled in the mid-1850s.[6]

However, probably the worst of these failures concerned the most spectacular refractor of the 19th century. This instrument (the largest ever made) had been planned by a private society for the 1900 Paris Universal Exhibition and was supposed to be one of the scientific marvels of the event (Launay 2007). Due to its focal length of 57 m (!), it would have been impossible to adopt an equatorial mounting and to build a dome for such a tube. Hence, it was decided that a horizontal fixed tube would be installed, while the image of the sky was reflected into the objective thanks to a gigantic siderostat (Fig. 12). This weighed 70 tons and had a 2-m diameter moveable glass mirror, floating on a mercury bath. It should

[6]A detailed history of Craig's telescope can be found in: http://www.craig-telescope.co.uk/site.html (accessed 25 October 2016).

Fig. 12 The tube with the photographic plate holder of the Paris 1900 exhibition telescope made by Gautier. The 2-m mirror of the siderostat is visible in the background. The image is a photomontage because the people were added later. [From: *Le Panorama*, Nouvelle série, N. 14, Paris 1900, pages unnumbered]

have had two interchangeable objectives (one visual and one photographic) of 125-cm diameter, but only one of them was completed. The telescope was certainly an incredible technological achievement, but, unfortunately, its location in the centre of Paris, where the atmosphere was polluted by the smoke of the chimneys and the lights of the city, was far from ideal. After the exhibition, the instrument was disassembled. Various projects, aimed at reinstalling it (for example, at the Pic-du-Midi Observatory) never concretized, and finally, in the 1920s, it was scrapped.[7]

The Paris telescope probably represented the swansong of the gigantic achromatic doublets. Some other large refractors were built in the very first decades of the 20th century, but none of them reached the apertures of the telescopes of the Lick, Yerkes and Meudon Observatories. A diameter of 1 m (or a little more) was the practical limit for the objective lenses. In fact, several factors prevented this size from being surpassed. Apart from the difficulties in producing such optical elements, larger and thin lenses risked being slightly deformed under their own weight[8] and, additionally, it was impossible to increase further their thickness because they would have absorbed too much light. By the end of the 19th century, the technology of large refractors had reached its peak. Various other large refractors were built in the first decades of the 20th century, but the golden era of these instruments was definitely ending.

[7]Only the large mirror (unsilvered) and the objective survive today in the collection of the Paris Observatory.

[8]The achromatic doublet of the Yerkes objective weighed more than 220 Kg.

6 The Decline of Refractors

During the 19th century, professional astronomers generally preferred large refractors to reflecting telescopes. Both instruments had their advantages and disadvantages. Refractors were high-precision instruments, ideal for traditional positional astronomy; furthermore, they did not require much attention. On the other hand, they were extremely expensive, and the difficulty of having large and homogeneous optical glass blanks limited the size of their objectives, which always showed a residual chromatic aberration. Reflector telescopes were cheaper to produce, because only one surface had to be figured, instead of the four surfaces of a typical achromatic doublet. Moreover, mirrors did not have any chromatic aberrations (residual chromatic aberrations were not a significant problem for visual observations, but were a handicap for photography). Indeed, early large metallic mirrors, which were made of speculum metal (an alloy of copper and tin), were difficult to cast; they also oxidized rapidly, and therefore had to be frequently polished. Such an operation was extremely delicate and could easily modify the curvature of the mirror, thus deteriorating the quality of the images. Only upon the introduction of silvered glass mirrors into the reflectors of the late 1850s, was a new era opened in the development of reflectors. Another disadvantage of early reflectors consisted in the fact that the mounting for a large and heavy mirror was difficult to manoeuvre and was unpractical for precision measurements. Finally, with the Newtonian reflecting telescopes, astronomers had to look in the eyepiece near the top of the tube, in an often uncomfortable and sometimes dangerous position. Therefore, barring a few exceptions, for the better part of the 19th century, reflectors were mainly used by wealthy amateur astronomers, such as William Parsons (1800–1867), third Earl of Rosse, who made and installed, in his Irish estate, the famous "Leviathan of Parsonstown", a gigantic reflector with a mirror 1.8 m in diameter and a tube about 16 m long. Of course, in the mid-19th century, Lord Rosse and other amateurs, using large reflectors, were more interested in exploring the deep universe than in collecting measurements for positional astronomy.

Only in the last decades of the 19th century, and with the introduction of better mountings, good silvered mirrors and several technical improvements, did reflecting telescopes with large apertures and great light-gathering power become serious competitors to refractors. These factors, together with the spectacular development of astronomical photography and spectroscopy, and the interest in exploring stars and nebulae, made reflectors the most important observational instruments of the 20th century.

Fig. 13 The Potsdam double refractor (visual and photographic) of 1899 made by Repsold with the objectives of Steinheil. [From: Repsold (1914), image related to p. 53]

7 "Big Is Beautiful": Yesterday and Today

If large refractors (or at least some of them) contributed greatly to the development of 19th century astronomy and astrophysics, they also certainly played an emblematic and political role. Having a well-equipped observatory with a large refractor (possibly the largest) was a matter of national pride and a powerful weapon in the battle for scientific supremacy. The Lick and Yerkes refractors represented the fast rise of American astronomy and the role of American tycoons in sponsoring science. The Potsdam double refractor, inaugurated in 1899 (Fig. 13), was held up as a symbol of the science of the German empire and was inaugurated with great pomp by Kaiser Wilhelm II. The Nice and Meudon refractors and the ill-fated 1900 exhibition telescope were intended to demonstrate French scientific supremacy and the excellence of its scientific instrument industry.

Furthermore, large telescopes fired peoples' imaginations. They were the instruments of the late 19th century's "big science". They promised astonishing discoveries. They allowed for exploring far away and mysterious worlds, where other forms of life perhaps existed. They were the symbols of a triumphant and positivistic science. Therefore, the public was fascinated by them, and they were often described and illustrated in popular magazines and journals.

Today, apart from a very few exceptions, the large refractors made in the 19th and early 20th centuries are no longer used. These technological giants of astronomy rest under their domes as important material witnesses to 19th century astronomy. Some of them have been restored in recent years, and from time to time are used for didactic observations. A few others have been reinstalled and displayed in museums. This last solution, which is sometimes necessary, is not ideal, because telescopes, together with their ancillary equipment and their buildings, form a single complex unit. To break it apart is like dismantling an instrument: it entails a loss of information, and the separated elements do not make much sense anymore.

However, preserving and restoring large refractors (and their domes) is very expensive, and these tasks are not priorities for contemporary astronomical and astrophysical institutes. For these reasons, the future of some of these telescopes is at risk. Therefore, public institutions, museums, historians of astronomy, amateur astronomers, and potential sponsors, as well as the public, have to consolidate their efforts in order to preserve a spectacular and fundamental part of our scientific and historical heritage.

References

AMBRONN, LEOPOLD: *Handbuch der Astronomischer Instrumentenkunde* (2 volumes), Berlin, Julius Springer, 1899.

BELL, LOUIS: *The telescope*, New York, Dover Publications, (reprint of 1922 original edition), 1981.

BRENNI, PAOLO: *An Italian in Paris: Ignazio Porro (1801–1975)*, unpublished manuscript.

CLARK, ALVAN and EDOUARD MANTOIS: "Les grands instruments de l'avenir", *L'Astronomie*, Mai 1894, pp. 13–31.

DANJON, ANDRÉ and ANDRÉ COUDER, *Lunettes et télescopes: théorie, conditions d'emploi, déscription, réglage,* Paris, Librairie Blanchard (reprint of the 1935 original edition), 1979.

DUVAL, H.J.: "The Greatest Telescope in the World", *Metropolitan Magazine*, Vol. III, N. 1, 1896, pp. 11–16.

HUSSEY, WILLIAM J.: "The Yerkes Observatory", *Publications of the Astronomical Society of the Pacific*, Vol. IX, N.59, 1897, pp. 209–222.

KING, HENRY C.: *The History of the Telescope*, New York, Dover Publications (reprint of 1955 original publication), 1979.

KOST, JÜRGEN: *Wissenschaftlicher Instrumentenbau der Firma Merz in München (1838–1932)* Hamburg, Tredition, 2015.

LAUNAY, FRANÇOISE: "The Great Paris Exhibition Telescope of 1900", *Journal for the History of Astronomy*, vol. 38, 2007, pp. 459–475.

LEQUEUX, JAMES: "The great nineteenth century refractors", *Experimental Astronomy*, 2009, vol. 25, pp. 43–61.

PORRO, IGNAZIO: *Catalogue raisonné des produits de l'Institut Technomatique*, Paris, chez l'auteur, 1857.

REPSOLD, JOHANN A.: *Zur Geschichte der astronomischen Messwerkzeugen von 1830 bis um 1900,* Leipzig, Emmanuel Reinicke, 1914.

RIEKHER, ROLF and HANS BECK: *Fernröhre und ihre Meister,* Jena, Hans. G. Beck, 2012, (Korrigierter Neudrucl der 2. Auflage von 1990).

STRUVE, *Beschreibung des auf der Sternwarte der Keiserlichen Universität zu Dorpat Befindlichen grossen Refractors von Fraunhofer*, Dorpat, J.C. Schürmann, 1825

STRUVE, FRIEDRICH GEORG WILHELM VON: "Nachricht von der Ankunft und Aufstellung des Refractors von Fraunhofer auf der Sternwarte der Kaiserl. Universität zu Dorpat", *Astronomische Nachrichten,* vol. 4, 1826, p. 37–44 and pp. 49–52.

VAN HELDEN, ALBERT: "Telescope Building 1850–1900", in Owen Gingerich ed. *The General history of astronomy, vol. 4, Astrophysics and twentieth century astronomy to 1950: Part A*, Cambridge, Cambridge University Press, 1985, pp. 40–152.

The Merz Company: A Global Player of 19th Century

Gudrun Wolfschmidt and Jürgen Kost

1 Joseph von Fraunhofer: A Sort of Hagiography

In the early 19th century, as a result of the Continental System, the Bavarian workshops achieved global leadership in high quality optical-mechanical instrument-making thanks to the work of Fraunhofer and his successors (Jackson 1993).

Joseph von Fraunhofer (1787–1826) began his career in 1806 at the Mathematical-mechanical Institute of Munich. In 1809, he became leader of the glassworks in Benediktbeuern (Seitz 1929), where he invented a device used to stir the glass melting in the furnace, in order to make it free of streaks (see Brenni, in this volume). He also developed methods to produce flint and crown glass of a given dispersion (first, by calculation of the dispersions, and second, by using the correct mixture of ingredients). With this expertise, he could make large high quality achromatic lenses, like the one for the Dorpat refractor (1824).

In 1821, he was appointed as an extraordinary member of the Bavarian Academy of Sciences and, two years later, professor and conservator of the physical cabinet of the Bavarian Academy of Sciences.

This brilliant ascent was interrupted in 1826, when Fraunhofer died, at the age of 39 years, in Munich.

The Institute of Mathematical Mechanics, founded in Munich in 1804 by Joseph von Utzschneider (1763–1840), Georg Reichenbach (1771–1826), and Joseph Liebherr (1767–1840), was one of the most important starting points for this hegemony. Most literature about Fraunhofer was written on special occasions, such as the jubilee of his birthday, the anniversary of his death or the inauguration of a

G. Wolfschmidt (✉)
Hamburg University, Hamburg, Germany
e-mail: gudrun.wolfschmidt@uni-hamburg.de

J. Kost
Tübingen University, Tübingen, Germany

I. Chinnici (ed.), *Merz Telescopes*, Historical & Cultural Astronomy,
DOI 10.1007/978-3-319-41486-7_2

monument or of the Benediktbeuern Museum. Some examples of the topic ... *aus Fraunhofers Hand* (... made by Fraunhofer) are highlighted below:

- *Der nach Pulkowa (bei Petersburg) gelieferte Refractor Fraunhofers [...]* (Preyß 2008, 72)

  (The Fraunhofer refractor delivered to Pulkowa (near St. Petersburg) [...])—Fraunhofer had already been dead from thirteen years, when the instrument was delivered, in 1839!
- *[...] es war ehrliches Bemühen tüchtiger Leute, – aber ohne Wissen um den Weg in die Zukunft und die Fähigkeit, auf Fraunhofers Fundament wissenschaftlich und technisch weiterzubauen.* (Preyß 2008)

  (Capable people tried hard, but without any knowledge about the path to the future and without the skill to build on Fraunhofer's fundament in a scientific and technological way).
- *[...] nie mehr so gute Instrumente gebaut wurden wie zu Fraunhofers Zeiten* (Sang 1987, 133)

  (Such good instruments, as were made in Fraunhofer's time, were never built anymore).
- *[...] unaufhaltsamen Niedergang dieses Betriebes nach Fraunhofers Tod.* (Sang 1987, 27)

(The inexorable decline of this workshop after Fraunhofer's death).

Fig. 1 Fraunhofer as a hero: Joseph von Fraunhofer demonstrates the spectroscope. *From the left* Joseph von Utzschneider, Joseph von Fraunhofer, Georg von Reichenbach, Joseph Liebherr (?) and Georg Merz (painting by R. Wimmer 1897)

This overestimation of Fraunhofer, in comparison to his successor Merz, is a trend of many biographers and a classic example of hagiography, as these historians completely neglect the importance of the Merz Workshop.

A famous painting by Rudolf Wimmer (1897) (Fig. 1) shows Joseph von Fraunhofer as a hero, while demonstrating his new spectroscope. You can also see Georg Merz here, relegated to the background of Wimmer's painting—despite the fact that he was Fraunhofer's successor and the new head of the glassworks!

It is a myth that this tradition in making optical instruments ended with Fraunhofer's death. A prominent feature, by which the company G. & S. Merz can be characterized, is their introduction of many technical innovations over a period lasting more than a century (from 1826 to 1932), thus achieving a leading position in scientific instrument-making.

2 The Merz Company—Sources and Traces

2.1 Sources and Methods of Research

Some questions should be considered as a starting point for research carried out about the Merz Company:

- How did the Company develop under Fraunhofer's successors—the Merz family?
- How successful was the Company and which seller's markets did they open up?
- What role did the instruments of the Merz Company play in the field of science?
- What was the reason for the final collapse of the firm?
- Why did the Merz Company disappear after being in operation for more than 100 years?

The sources and methods used for the research work on the Merz Company are listed below:

- Compilation of biographies of the important protagonists of the Optical Institute after 1826.
- Analysis of different inheritances, e.g., Sigmund Merz papers (Deutsches Museum Munich) and the collection of the optician Rudolf Loher (Stadtmuseum Munich, NL Loher) (Loher 1964).
- Looking for traces in Munich and BenediktbeuernBenediktbeuern (oral history).
- Three-dimensional sources—looking for instruments in museums and private collections.
- Compiling a register of instruments by considering the customers.
- Constructing a portfolio of the products of the Merz firm.

Many printed sources, such as company publications, trade literature, advertisements and publications in journals like *Astronomische Nachrichten* and *Zeitschrift für Instrumentenkunde*, etc., have also been used.

2.2 Trade Literature and Account Books

The Optical Institute and the Merz Company published schedules of prices (*Preiscourant, Verzeichniß der optischen Instrumente*) from 1811 to 1930.

The trade literature (*Firmenschriften*) comprises, as a whole, 36 advertising brochures (catalogues and technical communications from customers) from the years 1826 to 1932, which could be analysed. But the real portfolio of the products of the Merz firm was considerably larger![1]

The Account Books (business records) of the Merz Firm document the years from 1835 to 1881; there are 13 volumes, with about 1800 pages preserved. In these books, you can find, e.g.:

- Instrument descriptions, construction elements, etc.
- Information about customers
- Delivery dates of the instruments
- Payment arrangements
- Repair orders.

This was the most important source for the registry of instruments!

Fraunhofer's successors shaped the company significantly and opened up new sales markets. In addition, the large collection inherited by the historian of optics Rudolf Loher (1900–1975) is a very valuable source; it contains a large number of photographs, papers and instruments.[2]

In Munich, one can look for traces of the Merz firm. The residential Merz building and workshop at Müllerstraße 11 still exists (Fig. 2). Another important photograph shows how the Fraunhofer/Merz Workshop was arranged in an exhibit held at the City Museum (Stadtmuseum) in Munich (1964); one can recognize many original tools and instruments. Finally, there is the gravestone of the Merz family (Fig. 3) in the Old Southern Cemetery in Munich.

[1] About Merz catalogues of the years 1832, 1878, 1900, see: Kost 2015.

[2] For example, a filar micrometer (1880) and an objective-glass d = 38 cm (Bordeaux II).

Fig. 2 Residential building and Workshop of Merz at Müllerstraße 11 in Munich, around 1900 and in the year 2012. On the roof, the observatory for testing the instruments! (Stadtmuseum Munich [StAM, NL Loher]; *credit* Jürgen Kost)

Fig. 3 Family grave stone of Merz in the Old Southern Cemetery in Munich, around 1905 and in 2012. The right plaque in the *left* image is empty, awaiting the portrait of Sigmund Merz. (StAM, *credit* Kost)

3 Epochs of the Optical Institute and Merz Company

The epochs of the Optical Institute and the Merz Company are given below, from its establishment to the cessation of its activity, together with the name of the Company, which changed over the years.

- (1809)—Founding of the Optical Institute
- (1809–1826)—Fraunhofer (17 years)
- (1826–1839)—Utzschneider and Merz (13 years)
- (1839–1845)—Merz and Mahler (6 years)
- (1845–1858)—Merz and Söhne (13 years)
- (1858–1867)—Georg and Sigmund Merz (9 years)
- (1867–1883)—Sigmund Merz (16 years)
- (1883–1903)—Jakob and Matthias Merz (20 years)
- (1903–1932)—Paul Zschokke (29 years).

It is worth noting the brevity of the Fraunhofer epoch in comparison to the many years of the Merz Company (Fig. 4).

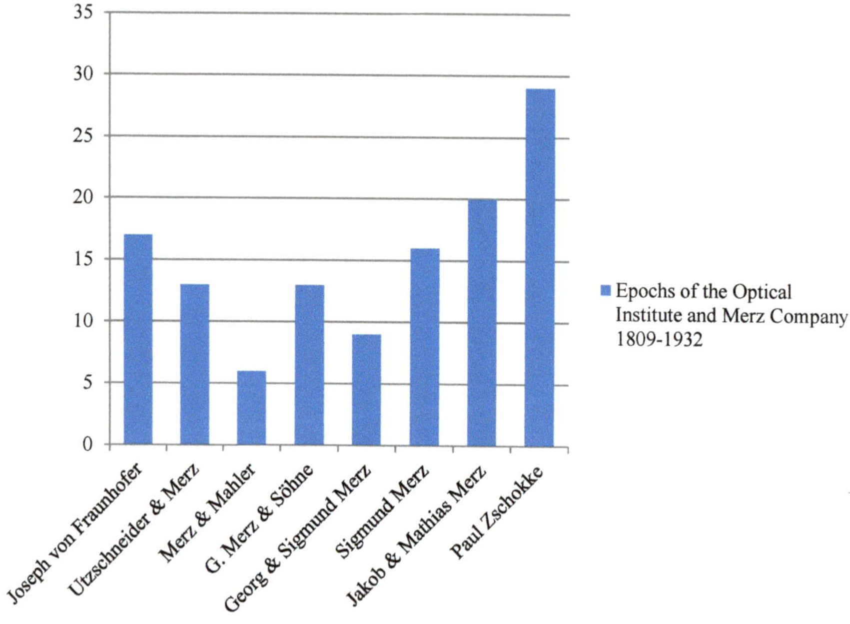

Fig. 4 Epochs of the Optical Institute and Merz Company, from 1826 to 1932. (*Graphics* Jürgen Kost)

3.1 *The Founders of the Merz Company: Georg Merz and Joseph Mahler*

After gaining considerable experience as a co-worker of Fraunhofer and, later, of Utzschneider, Georg Merz (1793–1867) founded the Merz workshop (Merz 1868) with the financial support of his friend Joseph Mahler (1795–1845).

The main products and activities of his workshop were:

- Large refractors and heliometers, very successful instruments at the beginning of the 19th century, including: the Bonn heliometer (1841); the 38-cm-aperture refractors at the Pulkovo Observatory (1839) and at the Cambridge Observatory (1847).
- Improvements of microscopes (i.e., the prismatic microscope, designed by G. Merz).
- Optics (spectacles, magnifying lenses, etc.).

Around 1858, the Merz workshop for glass grinding and polishing was located behind the *Viktualienmarkt* in Munich (Fig. 5a). Tools for the lens grinding machine and for polishing are kept today in the Fraunhofer Glashütte Museum in Benediktbeuern (Fig. 5b).

3.2 *The Period of Prosperity—Sigmund Merz: 1845–1883*

This period includes three epochs, during which the leadership of the company gradually passed from Georg Merz to other members of his family (Fig. 6), starting with his son, Sigmund (1824–1908):

(1845–1858)—G. Merz and Söhne
(1858–1867)—Georg and Sigmund Merz
(1867–1883)—Sigmund Merz.

(a) **(b)**

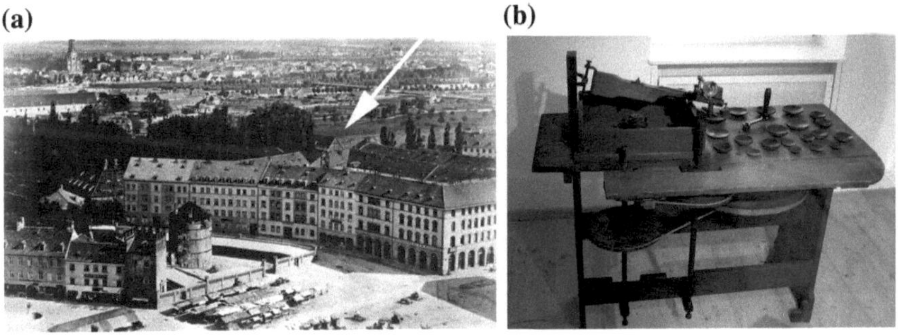

Fig. 5 **a** Merz workshop in Munich, after 1858. **b** A small lens grinding machine and polishing bowls, (Stadtmuseum/City Museum Munich [StAM], Museum Fraunhofer Glashütte in Benedikt-beuern)

(a)

(b)

(c)

(d)

(e)

(f)

◂**Fig. 6** The Merz family: **a** Georg Merz with his improved aplanatic microscope **b** Joseph Mahler (Private collection, Stadtarchiv Immenstadt). **c** Ludwig Merz (1817–1858), **d** Sigmund Merz, (Engraving by: A. Schleich, Bayerische Staatsbibliothek Munich). **e** Jakob Merz and **f** Matthias Merz, photography, around 1880, (Private collection, Stadtmuseum Munich [StAM, NL Loher])

The second workshop of the Merz Company was set up at Blumenstraße 31 in Munich, and was used from 1867 to 1907, up until the epoch of Jakob Merz.

The typical products of this period were:

- New refractors, including: the 24-cm aperture refractor at the Quito Observatory (1875); the 18-cm refractor at the Calcutta Observatory (1876).
- Improved microscopes for physicians, students, meat inspectors, etc., such as: the small Merz Microscope (1862); the Dissection Microscope (1868); the Trommel Microscope (1860).
- Military optics, such as: marine double binoculars (1885); Bavarian military telescope (1867); Marine double binoculars (1865); Marine double binoculars "Fürstenmodel"; distance-measuring instruments.
- Optics for photography (photography had become an emerging field for the public since the 1880s, spurred by the introduction of the dry photographic plate process).
- Spectroscopes and objective-prisms (e.g., 16-cm diameter objective-prism 1869). Merz began the first serialized production of astronomical spectroscopes, such as the large universal star spectroscope and the small universal star spectroscope (1910).
- Objectives and optical elements for many well-known workshops (Breithaupt of Kassel, Ertel of Munich, Repsold of Hamburg), as well as spectacles and glasses for Rodenstock Company of Munich.

One of Sigmund Merz's most important contributions was a list of glass-melting processes in the Benediktbeuern Glassworks from 1842 to 1885. This list of 431 glass meltings gives very good insight into the production process. There were 113 crown glasses in comparison to 318 flint glass meltings, thus giving a ratio of about 1 to 2.8.

3.3 *The Last Active Members of the Merz Company—Jakob and Matthias Merz: 1883–1903*

After the death of Sigmund Merz, the company passed into the hands of his brothers, Jakob (1833–1906) and Matthias (1826–1883), in collaboration with Paul Zschokke (1853–1932) (Fig. 7a) (Zschokke 1818).

Fig. 7 **a** Paul Zschokke, **b**: Zschokke Villa at, Hauptstraße 21, Munich-Pasing in 1902, (Zschokke family archives, Flüh/Schweiz, Kost, 2013/StAM)

Their main products and activities were:

- Scholastic and amateur telescopes for the educated class (Bildungsbürgertum).
- Repairing and maintenance of large telescopes.

3.4 The Final Epoch—Merz-Zschokke: 1903–1932

In the years from 1903–1932, the company was renamed G. & S. Merz G.m.b.H.—*G. & S. Merz, Optisches Institut, ehemals Utzschneider & Fraunhofer*. In the catalogue from 1911, 332 instruments were listed.

The main product was:

- Schupmann Medial telescope (Fig. 8), mainly for amateur astronomers.

There exists a Merz advertising brochure for this instrument, *Das Medial-Fernrohr* (1924), which is known in English-speaking countries as the Schupmann telescope (Schupmann 1913). Merz also built a successful 38.5-cm aperture medial telescope for the private Landstuhl Observatory (1912). Only amateur astronomers were interested in the medial telescope, but it was too expensive for many of them. The professional astronomers disliked its catadioptric design, which was a cross between a refractor and a reflector.

(a) (b)

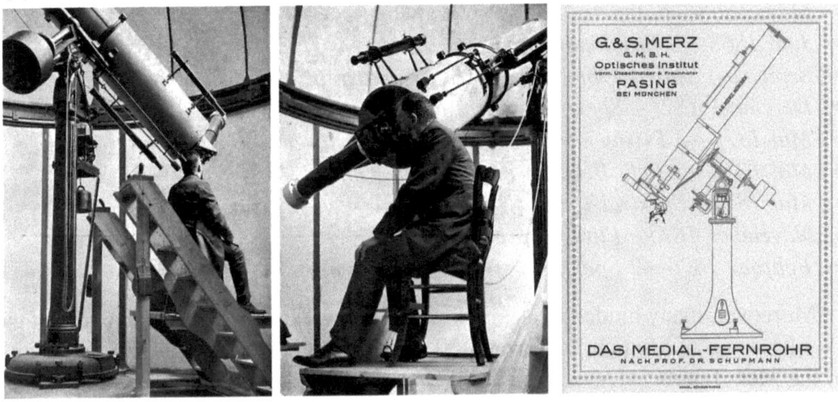

Fig. 8 **a** Medial telescope, used by Philipp Johann Heinrich Fauth (1867–1941), in his private observatory at Landstuhl, Palatinate, later Munich-Grünwald. **b** Medial Merz Telescope advertising flyer (1924), (From: Merz-Medialfernrohr Werbeschrift 1924, Fauth P. 1912, Tafel 1)

3.5 *Staff of the Merz Company*—Invisible Hands

It would be very complicated to compile a complete list of the company's employees, including opticians, precision mechanics and assistants (Hentschel 2008); however, a list of the staff of the Merz firm in 1883 has recently been published (Kost 2015). One of the most important employees was Rudolph Weiss (1809–1882), who worked at the Merz Company for 54 years, acquiring good expertise in optical production.

4 Data from Account Books and Other Sources

From the Account Books, a list of refracting telescopes, objectives and lens tubes made by Merz has been compiled (Kost 2015); the list includes 43 refractors and 93 objectives/lens tubes.

Additional information can be extracted from this source; for example, here is a list of all the Merz instruments delivered to Angelo Secchi (1818–1878), many of which are devices for spectroscopy and solar physics:

- *Januar 1853—2 Doppelring-Mikrometer*
- *August 1854—9—zölliger Merz-Refractor, Beugungs-Apparat*
- *Juni 1856—Mikrometerokular zum 9 Zöller*
- *März 1863—Zylinderlinsen mit 13'''Öffnung*
- *Oktober 1865—Spektralapparat mit Doppelprismen 24''''*
- *April 1866—Kompensations-Prisma for den Spektral-Apparat*

- *Oktober 1866—Zylinderlinse 12'''Öffnung*
- *März 1868—Kompensations-Prisma 8°*
- *Juni 1868—Schraubenmikrometer*
- *September 1869—Objective-Spektral-Apparat 6''Öffnung*
- *Dezember 1871—Spektralapparat Nr. 80*
- *Mai 1872—2 Prismen 90°, 12 Stück Sonnengläser*
- *Mai 1873—2 Refl. Prismen und 2 kleine Objective*
- *Mai 1874—Spektral-Prisma 18–33 mm, 1 Rautenprisma*
- *November 1875—Flintglasprisma 60° 22'''*
- *Februar 1876—2 Spektral-Prismen 18–24 mm.*

Moreover, many hidden treasures were found by Kost in 2015 in the depot of the City Museum of Munich, including 100 engineering drawings from the inheritance of the Merz family (StAM, Sammlung Grafik 32/168 ff.) which will be analysed in the near future.

Archival materials related to the Merz company have been found at:

- The Carl Zeiss Archive, Jena
- The Archive and Collections of the Deutsches Museum in München
- The Dr. Johannes Heidenhain GmbH in Traunreut
- The Institute for Astronomy and Astrophysics of Tübingen University
- The Stadtmuseum (Citymuseum), Munich
- The Stadtarchiv Oberstaufen
- The SAO/NASA Astrophysics Data System
- The Tübingen University Archive
- Private collections and archives.

5 Merz as a Global Player—Different Types of Instrument

In particular, the analysis of the account books and the correspondence of the Merz Company, which have remained unscrutinized up to now, reveal the company's extensive economic ties as a supplier of lenses and optics for no less than 30 firms in Germany and abroad (Figs. 9 and 12).

Below is a list of the main types of instrument manufactured by Merz (Kost 2015).

Objectives and Spectroscopic devices (Fig. 10)

- 12" objective lens
- Objective prism (c.f. Collegio Romano 1869, Berlin Observatory, Dun EchtDun Echt Observatory 1873)
- Large Merz-Universal-Star Spectroscope Nr. 78

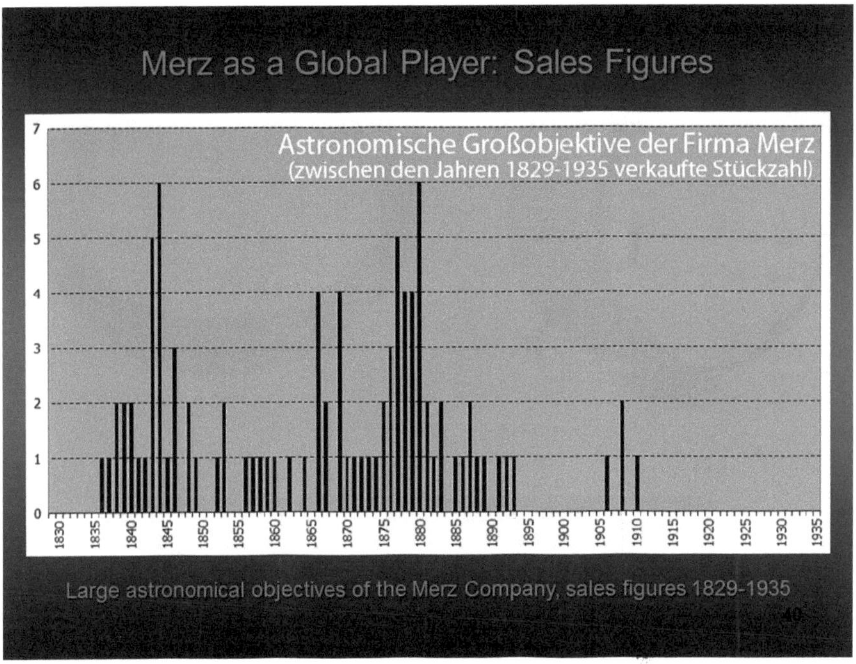

Fig. 9 Large astronomical objectives of the Merz Company, sales figures 1829–1935 (*Graphics* Jürgen Kost)

- Spectroscope, sold to Steinheil
- Merz/Cooke 7" Refractor with a 5-prism spectroscope, (Kalocsa Observatory 1878).
- Helioscope for solar physics (1876)

Transit Instruments, Meridian and Vertical Circles

- Repsold/Merz Meridian Circle, Pulkovo Observatory, 1838
- Meridian Circle, Spain, 1852
- Meridian Circle, Strasbourg/Straßburg, 1875 + 1877
- Meridian Circle, Tokyo, 1878
- Meridian Circle, Munich, 1891
- Meridian Circle, Bonn, 1893
- Repsold/Merz Transit Instrument, Pulkovo Observatory, 1834/39
- Repsold/Merz Vertical Circle, Pulkovo Observatory, 1834/39

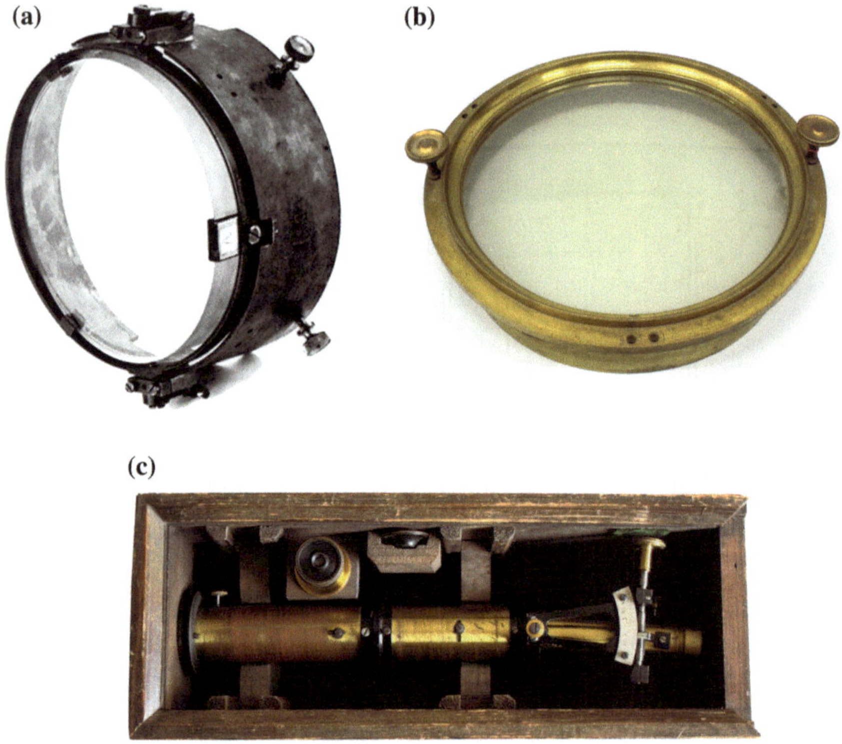

Fig. 10 Astronomical optics, made by Merz: **a** Objective prism of Collegio Romano Observatory, 1869 (INAF-OAR), **b** 12" objective with brass mounting (StAm, NL Loher), **c** small star small star spectroscope, around 1910 (collection Rolf Riekher, Berlin)

Heliometers

- Königsberg Heliometer (Kaliningrad, Russia), 1829. (It was commissioned by Friedrich Wilhelm Bessel (1784–1846). Fraunhofer secured the contract and created the glass-melting procedure. Merz took up the project in 1827, immediately after Fraunhofer's death, effecting the cutting of the lenses, which was a very complicated operation).
- Bonn Heliometer, 1841 (restored in 1893) (Fig. 11a)
- Radcliffe Observatory Heliometer, Oxford, 1844
- Yale College 6-inch Heliometer, New Haven, Connecticut, 1880
- Capetown Observatory Merz/Repsold 7-inch Heliometer, 1887—the largest heliometer in the Southern hemisphere.
- Remeis Observatory Repsold/Merz 7-inch Heliometer, Bamberg, 1889—the largest heliometer in the Northern hemisphere.
- Munich Heliometer, 1891

(a) (b)

Fig. 11 Merz Instruments: Heliometer and Equatorial. **a** 6" Merz Heliometer of the Bonn Observatory, 1841, (restored in 1898). **b** Repsold/Merz Equatorial, Hamburg, old Millerntor Observatory, 1864 (now in Bergedorf), (Deutsches Museum Bonn, Hamburg Observatory)

Equatorial telescopes (Kost 2011)

- Millerntor Observatory 9-inch Repsold/Merz Equatorial, Hamburg, 1867 (this was a special new type of instrument, to be used in place of meridian circles for compiling precise star catalogues) (Fig. 11b)
- Christiania/Oslo Observatory Equatorial, 1842
- Altona Observatory Equatorial, 1858
- Gotha Observatory Equatorial, 1860

Astrographs

- 9-inch Astrograph with Micrometer, Kiev Observatory (an astrograph is a special double refractor for photographing the sky; one lens is corrected for the blue wavelength range—as early photographic plates were only sensitive in that range—and the other for visual observations).

Refracting Telescopes (Fig. 13)

- 14-inch refractor, Harvard College, Cambridge, Mass., 1847
- 9-inch refractor, Collegio Romano, 1854

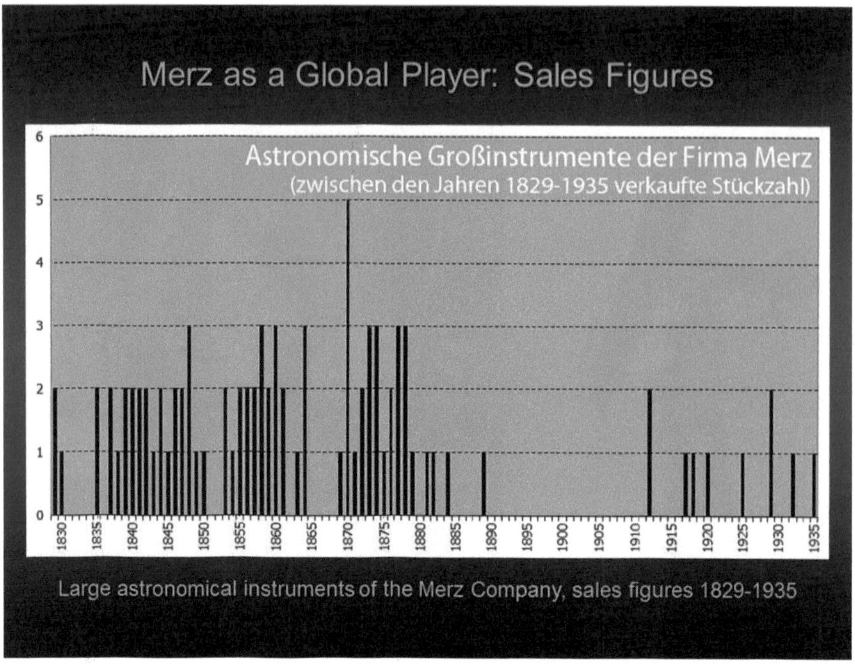

Fig. 12 Large astronomical instruments of the Merz Company, sales figures 1829–1935 (*Graphics* Jürgen Kost)

- Lisbon Observatory, 1861
- Santiago Observatory, 1869
- Nikolajew Observatory, Kiev, 1874
- Stockholm Observatory, 1876
- 18-inch Repsold/Merz refractor, Strasbourg Observatory, 1870s
- Altazimut Refractor, Strasbourg Observatory, 1877
- Berlin Observatory, 1880
- Milan Observatory, 1881

Comet Seekers (Fig. 14)

- Göttingen Observatory, 1870
- Merz/Repsold Comet Seeker, Bamberg Observatory, 1889

Mass Production of Telescopes

- 6-inch refractor, World Fair in Chicago, 1893
- School Telescope, 1910
- Reflecting Telescope *G. & S. Merz*, 1920

Fig. 13 Merz Refracting telescopes, **a** 10½" Telescope, Munich-Bogenhausen Observatory, 1835. **b** 9" Telescope, O'Gyalla Observatory, Hungary (now Hurbanovo, Slowakia): the Zeiss astrograph, on the left, was added in 1904 (Leiden Observatory Archives)

Fig. 14 Merz Instruments for Amateurs: Comet Seekers and Reflectors, **a** Merz/Repsold comet seeker of the Bamberg Observatory with the so-called comet villa, around 1900. **b** Reflecting Telescope Newtonian type, G. & S. Merz, 1920, (Picture postcard, Merz advertising flyer, around 1920)

- 6-inch telescope for street astronomers, 1880 (in comparison with people who presented electrical, optical or chemical experiments to market places).

Between 1839 and 1932, the Merz Company delivered worldwide (Fig. 15), to practically all notable observatories, astro-optics, refracting telescopes, heliometers, etc. (Riekher 1990).

They not only produced astronomical instruments (meridian circles, transit instruments, heliometers, comet seekers, astrographs), but also military optics, surveying instruments, microscopes, and spectroscopes.

Merz produced approximately 90 large refractors and about 100 astronomical lenses. They were sold in England, France, Italy, the Americas, Russia, South Africa, India, the Philippines, Japan, and Australia. A list of these instruments is given in this book in Appendix B.

The Merz Company and their instruments made crucial contributions to the solution of many scientific problems, especially in the fields of astrometry, astrophysics, and geodetic instruments (theodolits), although bacteriology and medical diagnostics also deserve to be mentioned. Merz produced microscopes for biology and cristallography (Merz 1844) and also for medical purposes; Robert Koch (1843–1910) and Max Josef Pettenkofer (1818–1901) both used Merz instruments for their research. Microscopes continued to be delivered by Merz until roughly the 1880s, when more sophisticated microscopes began to be constructed by Leitz of

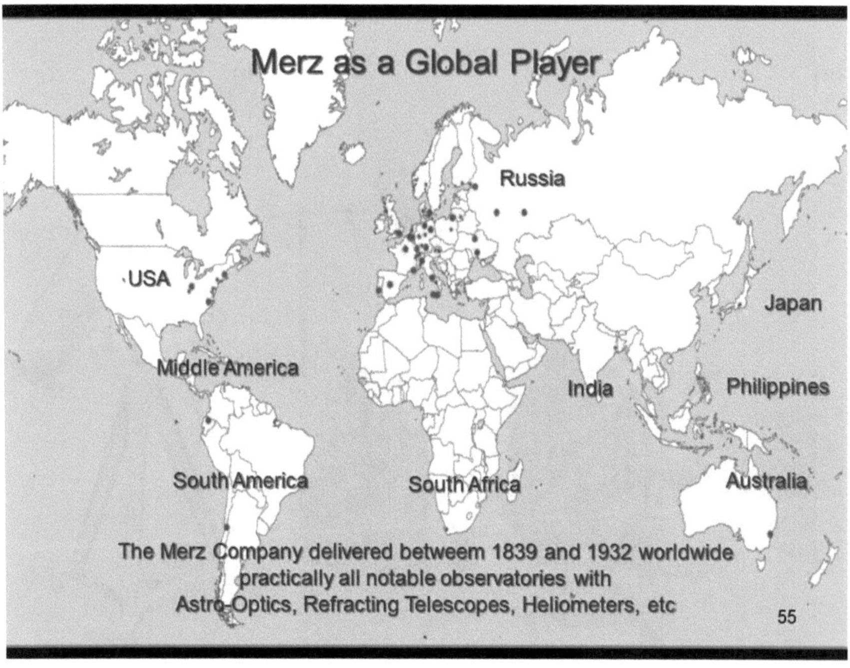

Fig. 15 Merz as a Global Player—world map of sold instruments, (*Graphics* Jürgen Kost)

Wetzlar (founded in 1869), Fuess of Berlin (founded in 1865), Hartnack of Potsdam (founded in 1864), Winkel of Göttingen (founded in 1857) and especially by Carl Zeiss of Jena (founded in 1846).

6 Why Did the Merz Company not Survive?

There are several reasons, which, when combined, could explain the demise of the Merz Company. They are listed below.

- Missed opportunities: The company failed to extend the important sector of photo optics and prism spy glass (Prismen-Feldstecher) going into the 1920s.
- Merz did not like to employ people from abroad, maintaining a strict policy of only hiring Bavarians.
- Merz only sold instruments out of their office in Munich, missing the opportunity to build up a dealer network like Carl Zeiss and Schott of Jena and Emil Busch of Rathenow did, with offices not only in other parts of Europe, but also in the USA, South America and Japan.
- New competitors began to spring up around 1900, e.g., Zeiss of Jena or Askania of Berlin.
- Paul Zschokke was not open to new developments like the medial telescope. Thus, he missed the opportunity to properly advertise this instrument. In addition, when Zschokke died in 1932, he had not prepared a successor.

6.1 The Output of the Merz Company

Sigmund Merz stated: *Ohne ein Fraunhofer'sches oder Merz'sches Objectivglas irgendwelcher Dimensionen wird es wohl keine Sternwarte geben* (Without a Fraunhofer or a Merz objective lens of any dimension, there will be no observatory).

The Merz Company was an extremely successful organization during the more than 100 years that they existed. Their impressive market areas characterize Merz as one of the early global players in the field of scientific instruments, especially in building of optical-mechanical instruments of high precision. Its influence on scientific instrument-making by far exceeded Fraunhofer's work.

"Sie brachten uns die Sterne näher!"

(They brought us nearer to the stars!)

References

HENTSCHEL, KLAUS: *Unsichtbare Hände*. Diepholz, Stuttgart, Berlin: GNT Verlag 2008.

JACKSON, MYLES W.: Die britische Antwort auf Fraunhofer und die deutsche Hegemonie in der Optik. In: *Deutsches Museum: Wissenschaftliches Jahrbuch 1992–1993* (1993), S. 117–138.

KOST, JÜRGEN: Die Äquatoreale der Firma Repsold in Hamburg. In: WOLFSCHMIDT, GUDRUN (Hg.): *Hamburgs Geschichte einmal anders – Entwicklung der Naturwissenschaften, Medizin und Technik, Teil 3*. Hamburg: tredition science (Nuncius Hamburgensis – Beiträge zur Geschichte der Naturwissenschaften; Bd. 20) 2011, S. 62–77.

KOST, JÜRGEN: Heliometer von Merz. In: WOLFSCHMIDT, GUDRUN (Hg.): *Von den Anfängen bis zur modernen Astrophysik. 125 Jahre Dr. Remeis-Sternwarte Bamberg (1889)*. Proceedings der Tagung des Arbeitskreises Astronomiegeschichte in der Astronomischen Gesellschaft 2014. Hamburg: tredition (Nuncius Hamburgensis – Beiträge zur Geschichte der Naturwissenschaften; Bd. 31) 2015, S. 330–345.

KOST, JÜRGEN: *Wissenschaftlicher Instrumentenbau der Firma Merz in München (1838–1932)*. Bearbeitet und herausgegeben von GUDRUN WOLFSCHMIDT. Hamburg: tredition (Nuncius Hamburgensis – Beiträge zur Geschichte der Naturwissenschaften; Band 40) 2015.

LOHER, RUDOLF: Die große Schau. In: *Foto-Magazin, Sonderdruck* 4 (1964), S. 82–84.

MERZ, LUDWIG: *Die neueren Verbesserungen am Microscope nebst den sie begleitenden Aenderungen in der Dioptrik*. München: Palm 1844.

MERZ, SIGMUND: Kurzer Lebensabriss von Georg Merz. Mitgetheilt von Herrn Dr. Sigmund Merz. In: *Astronomische Nachrichten* 70 (1868), Nr. 1679, S. 361–364.

PREYẞ, CARL R.: Joseph von Fraunhofer, Physiker und Industriepionier. München 2008.

RIEKHER, ROLF: *Fernrohre und ihre Meister*. Berlin: Verlag Technik (2. vollständ. überarb. Auflage) 1990.

SANG, HANS-PETER: *Joseph von Fraunhofer – Forscher, Erfinder, Unternehmer*. München: Verlag Dr. Peter Glas 1987.

SCHUPMANN, LUDWIG: Das Medial-Fernrohr zu Landstuhl. In: *Astronomische Nachrichten* 196 (1913), Nr. 4686, S. 101–106.

SEITZ, ADOLF: Die Utzschneider-Fraunhofersche optische Werkstätte nach Fraunhofers Tode und das Leben Georg Merzs des ersten Besitzers der Anstalt Utzschneider. In: *Deutsche Optische Wochenschrift* 5, 6, 7, 8 (1929), S. 55–59, 72–75, 87–89, 105–109.

WOLFSCHMIDT, GUDRUN (Hg.): Von den Anfängen bis zur modernen Astrophysik. 125 Jahre Dr. Remeis-Sternwarte Bamberg (1889). Proceedings der Tagung des Arbeitskreises Astronomiegeschichte in der Astronomischen Gesellschaft 2014. Hamburg: tredition (Nuncius Hamburgensis Beiträge zur Geschichte der Naturwissenschaften; Bd. 31) 2015.

ZSCHOKKE, H.: Die Werkstätten in Benediktbeuern, insbesondere das optische Institut. In: Gilberts *Annalen der Physik* 29 (1818), S. 196–205.

The Maker and the Scientist: The Merz-Secchi Connection

Ileana Chinnici

The Merz Company is known as one of the most successful instrument-makers in the nineteenth century (see Wolfschmidt and Kost, in this volume), but their interaction with scientists who commissioned their instruments is still largely unexplored. From a first examination of the exchanges they had with the Jesuit astronomer Angelo Secchi (1818–1878) (Fig. 1), it appears quite clear that the Merzs did not merely build instruments on demand: they actively collaborated in the scientist's research.[1] They were not simple laborers, but real partners, who gave suggestions, conducted tests, etc. Their contribution should not be neglected, especially in the case of the early spectral classification studies, mostly carried out by Secchi with Merz instruments.

The relationship between Angelo Secchi and the instrument-makers Georg (1793–1867) and Sigmund Merz (1824–1908) certainly dates back to at least 1852, when the Jesuit astronomer commissioned a large refractor (24.4-cm aperture, the largest in Italy at that time) from the German firm (see Altamore, Poppi and Maffeo, in this volume). The analysis of their correspondence (Fig. 2) reveals the role played by the Merzs in the evolution of Secchi's well-known early astrophysical studies, but also many other lesser-known aspects regarding the lives and activities of both the instrument-makers and the scientist. The part of the correspondence, which is here examined, consists of 71 unedited letters, dating back to the years 1856–1876 and sent by Merz to Secchi. The letters are kept at the Gregorian University in Rome, where an extended archival fund, containing Secchi's personal papers, is currently preserved. The Merz Company was Secchi's main correspondent,

[1]The same kind of collaboration seems to have been established by the Merzs with other scientists, as in the case of Johann Karl Friedrich Zöllner (1834–1882): see, for instance: Merz 1873a.

I. Chinnici (✉)
INAF-Osservatorio Astronomico di Palermo, Palermo, Italy
e-mail: chinnici@astropa.inaf.it

I. Chinnici (ed.), *Merz Telescopes*, Historical & Cultural Astronomy,
DOI 10.1007/978-3-319-41486-7_3

Fig. 1 Portrait of Angelo Secchi (from: *Annali*... 1879, p. [1])

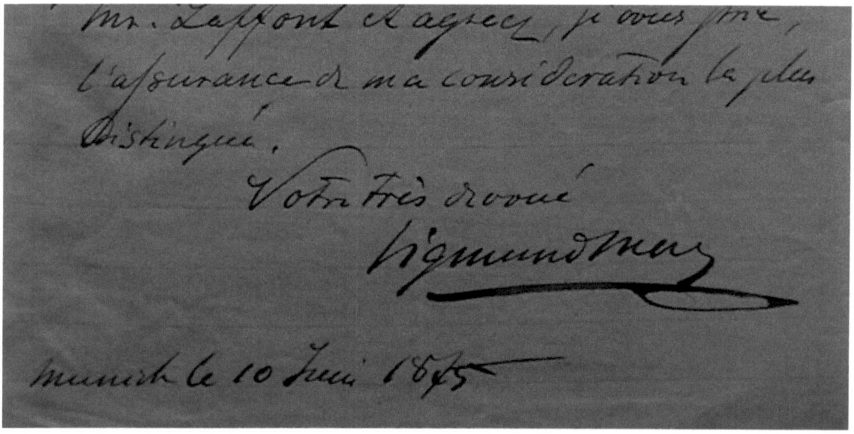
Mr. Laffont et agreez, je vous prie,
l'assurance de ma considération la plus
distingué.
Votre très dévoué
Sigmund Merz
Munich le 10 Juni

Fig. 2 Detail of the signature in a letter by Sigmund Merz (courtesy of INAF-OAR)

among the instrument-makers, and their letters are so detailed as to permit a good reconstruction of the whole content of their exchanges with the astronomer.[2]

1 Scientific Collaboration

When Secchi and Georg Merz began their collaboration, the Jesuit was Director of the Collegio Romano Observatory from a few years, and had already moved the Observatory from to a new location, on the roof of St. Ignatius' Church in Rome (Fig. 3). Georg had taken over the famous optical firm that Joseph von Fraunhofer (1787–1826) and Joseph Utzschneider (1763–1840) had established in Munich, and was continuing the renowned tradition of their optical institute. The Merz-Secchi correspondence increased in the 1860s, when the Jesuit astronomer carried out his studies on the spectral analysis of starlight. He could count upon an excellent team, having Merz in Munich as entrusted maker, who provided the instruments, and the brothers Emilio and Ermanno Brassart in Rome, as skillful technicians, who made adjustments and accessories.

In the context of the Merz-Secchi collaboration, the German opticians seem to have played a significant role in the dissemination of the Jesuit's devices. They often requested information about new improvements, made by Secchi to one or another apparatus, in order to apply them to the instruments they were constructing; Secchi, on the other hand, used to send suggestions to the Merzs, by his own initiative. In 1856, for example, Georg Merz wrote to Secchi that he had known, from some astronomers in Madrid, about a new kind of micrometer, proposed by Secchi (1855),

[2]Recent studies of the Merz archives; see: Wolfschmidt and Kost in this volume could lead to a complete reconstruction of the Merz-Secchi correspondence, in the future.

Fig. 3 Cross section of the Collegio Romano Observatory; on the right, the dome with the Merz equatorial (from: Secchi 1856)

and asked him for some details about this instrument, as "*it seems to be very advantageous for the science*" (Merz 1856).[3] In March 1865, the Merzs were working on some eye-pieces that were deviced by Secchi (Merz 1865a) and, one year later, while they was constructing new telescopes for the Observatories of Lisbon, Moscow and Montevideo, Sigmund asked Secchi for the permission to equip those telescopes with a new kind of spectroscope that the Jesuit had designed:

> Regarding your new spectroscope, I am grateful to you and I ask for your permission in order to construct it for our astronomical telescopes… […] I have also constructed the large spectroscope following the suggestions you gave me in June.[4] (Merz 1866b)

In 1868, Sigmund Merz made a Secchi-type solar spectroscope for the Moscow Observatory (Merz 1868f) and, three years later, he constructed nine spectroscopes of the same type (Merz 1871) for other buyers. Moreover, Sigmund indirectly participated in certain scientific debates between Secchi and other astronomers. In

[3]*… ci pare, che egli sia molto vantaggioso per la scienza.* Diplomatic transcription from the original letters has been adopted for all the quotations hereinafter. All the quotations from the letters will be reported in diplomatic transcription.

[4]*A ce qui concerne maintenant votre nouveau spectroscope je vous suis bien obligé et je profiterai volontairement de votre permission de le faire exécuter pour nos lunettes astronomiques* […] *Le grand spectroscope j'avais reconstruit aussi m'étant servi de vos conseils du mois de Juin.*

1873, for example, when Secchi was in dispute with Giuseppe Lorenzoni (1843–1914), astronomer at the Padua Observatory, about the best method for observing the transit of Venus with a spectroscope, Secchi involved Sigmund in the construction of an appropriate equipment for this kind of observations:

> … I have begun to carry out some spectroscopic experiences and I am ready to make the spectroscope, following the system that is proposed by you. I like this system very much and I hope that these instruments will provide better results than direct-vision prisms, in which the glue is always altered by the temperature. Regarding the price of such a double-path spectroscope, I think that it is possible to make it of the same size as my three-extra-dense flint-prism spectroscope No. 80, at 700 francs. The dispersion will be the same as that of your four-ordinary flint-prism instrument.[5] (Merz 1873b)

Another aspect of this scientific cooperation was Secchi's readiness to test Merz instruments. In 1863, during a visit of Otto Wilhelm von Struve (1819–1905) to the Collegio Romano Observatory, Secchi and the Russian astronomer decided to compare two eye-pieces, respectively made by Merz and the English astronomer Warren de la Rue (1815–1889). Once informed about this test, Sigmund Merz sent a newly constructed eye-piece to be examined too:

> I thank you very much for accepting my eye-piece and testing its efficiency in comparison with the one devised by Mr. Warren de la Rue […] Since you say that, together with Mr. Struve, you will choose the best of the two eye-pieces, Mr. De la Rue's and mine, I will take advantage of this noble competition and beg you to test a new eye-piece, just constructed, which I will send you immediately.[6] (Merz 1863b)

Again, in 1864, Sigmund asked Secchi to test a new kind of eye-piece for solar observations, before constructing another one, that was destined for the Pulkovo Observatory:

> Now I beg you, Mr. Director, to accept it and to let me know the result of its application to your refractor. I would be grateful to you, since I intend to provide Pulkovo Observatory with such an instrument.[7] (Merz 1864d)

---

[5] *… j'ai commencé des expériences spectroscopiques et j'entrera avec plaisir dans la fabrication su spectroscope selon le système que vous proposez. Le système me plait beaucoup et j'espère qu'en résultent des appareils beaucoup plus parfaits qu'avec les prismes à vision directe, dont la colle se change toujours par la température. Quant au prix d'un tel spectroscope à double trajet, je pense qu'on puisse exécuter l'instrument en dimensions de mon spectroscope N. 80 à 3 prismes en flint extra dense à fr. 700. La dispersion en serait la même comme dans votre instrument à quatre prismes en flint ordinaire.*

[6] *Je vous remercie infiniment de votre bonté de vouloir bien accepter mon oculaire et d'en épreuver la force en comparaison du système de Mr. Warren de la Rue. […] Lorsque vous me dites que vous choisirez avec Mr. Struve entre mon oculaire et celui de Mr. De la Rue je voudrais, profitant de l'occasion d'un si noble concours, vous prier de nouveau d' éprouver encore la valeur d'un oculaire, que je viens de construire et qui vous sera envoyé immédiatement.*

[7] *Je vous prie maintenant, Monsieur le Directeur, de vouloir bien l'accepter et d'en me communiquer le résultat en l'appliquant au réfracteur. Je vous serais bien obligé de votre complaisance ayant encore en vue de fournir aussi l'observatoire de Poulkowa d'un tel appareil.* The eye-piece was similar to the one described by Richard Hodgson (1804–1872) in 1854 (Hodgson 1854).

Sometimes, the Merzs also expressed their advice on certain scientific questions; for example, Georg argued that the controversial existence of a lunar atmosphere would have been demonstrated by the alteration of the solar spectrum during the contacts of an eclipse (Merz 1866c) and hypothesized the existence of different kinds of nebulae, having different spectral features:

> What you say about the nebula, that you have examined with the simple spectroscope, is really intriguing. I believe, however, that the slit should not be missing in the complete spectroscope. There are perhaps some nebulae which would stand as exceptions.[8](Merz 1866c)

The Merzs often gave information or suggestions to Secchi about devices or adjustments, as in the case of the heliometer:

> The simplest heliometric device is undoubtedly the heliometric eye-piece by Mr. Steinheil, which is mentioned by Mr. Struve in his description of the Pulkovo Observatory (page 195) [Struve 1845]. I believe, however, that your adjustment, by placing a small half lens between the objective-glass and the eyepiece of the large telescope, will be very advantageous. Mr. Lamont [*see below*], for example, for the same purpose, uses a prism with a very large angle, which is placed between the objective-glass and the eye-piece of the refractor. You would obtain the same result by using a rectangular prism, like that of the Amici meridian telescope, to be placed in the eye-piece.[9] (Merz 1874a)

On the other hand, the German makers often appealed to Secchi to address their doubts and get explanations, as in the case of the 1868 eclipse, visible from India, when solar prominences were spectroscopically observed for the first time by some international expeditions, causing Sigmund to wonder how the slit of the spectroscope was positioned to block the light coming from the solar disk (Merz 1868f).

2 Trading Aspects

The collaboration between Merz and Secchi was not limited to the scientific plane but also involved trading aspects. In 1862, when Sigmund Merz was able to make large objective-glasses, the Merz Company decided to sell the medium-size lenses at half-price and asked Secchi to publicize the offer (Merz 1862a).

---

[8] *Ce que vous dites de la nebuleuse examinée en employant le spectroscope simplifié est vraiment curieux. Mais je crois donc que la fente ne devrait manquer au spectroscope complet. Il y a peut-etre des nebuleuses qui en font exception.* On the early spectroscopic studies of nebulae, see: Huggins 1864.

[9] *Le plus simple appareil héliométrique est sans doute l'oculaire héliométrique de Mr. Steinheil dont Mr. Struve parle dans sa description de l'Observatoire de Poulkowa (page 195). Mais je crois que votre arrangement d'installer un petit objectif coupé en deux entre l'objectif et l'oculaire du grand téléscope vous conviendra fort bien. Mr. Lamont p. e. se sert pour le même bût d'un prisme à très grand angle placé entre l'objectif et l'oculaire du Refracteur. Le même resultat vous donnerait aussi un prisme rectangulaire comme celui de la lunette méridienne d'Amici si l'on voulait poser dans l'oculaire.*

Moreover, Secchi acted as a broker in obtaining optical instruments from Merz for other buyers: he received several microscopes (Merz 1864a), as well as small telescopes for amateur astronomers in Rome, usually noblemen like Princes Aldobrandini (Merz 1864b) and Orsini (Merz 1866b) and Count Barsanti (Merz 1870b), or churchmen like Cardinals Castellani, Lancellotti, Milori, and Cesi (Merz 1864c).

Sometimes, Merz solicited Secchi to publish the results of his observations—carried out with instruments made by the German opticians—in the most renowned astronomical journals of that time (Merz 1865a): this was, of course, a manner of "advertising" Merz instruments. Secchi never balked at doing this, thus contributing to the promotion of the German firm and the consolidation of its prestige.

On the other side, Merz did not miss any opportunity to circulate Secchi's scientific results; for example, in 1865, Sigmund informed the Jesuit astronomer in Rome that he had transmitted to his godfather, the physicist and astronomer Johann von Lamont (1805–1879), a photographic copy of a sunspot drawing that he had received from Secchi (Merz 1866a).

The Merzs rightfully considered themselves Fraunhofer's successors and the keepers of his tradition in optics. Sigmund was very proud of this heritage and desirous of recognition: in 1867, he donated a small telescope to Pius IX (1792–1878), hoping to be awarded, like his father Georg, with the Papal decoration of the Order of St. Sylvester. "*May the interest of the ancient optical institute by Fraunhofer justify my arrogance!*"[10] (Merz 1867a), he wrote to Secchi, looking for his support in this circumstance. Actually, the telescope donated by Merz was presented to the Pope by two of Secchi's assistants, the Jesuits Stanislao Ferrari (1834–1903) and Nazareno Mancini (1822–1870), who explained the usage of the instrument to Pius IX: "*He* [the Pope] *welcomed us very benevolently, we showed him the special improvements of your instrument, he wanted to practice with it by himself.*"[11] (Sigmund 1867b). The strategy was successful: the Pope rewarded Sigmund Merz for his gift by conferring upon him the desired distinction.

Merz also often appealed to Secchi's support and authority for facilitating relationships with other scientists. For example, on the occasion of the Universal Exhibition held in Paris in 1867, Urbain Leverrier (1811–1877), Director of the Paris Observatory, who was well known for being intractable, refused to certify the quality of the object-glasses that were sent by Merz for the Exhibition, and Merz turned to Secchi's mediation to overcome the problem (Merz 1867c).

Secchi undeniably contributed to extend the distribution of Merz telescopes all around the world. He often recommended Merz instruments for many Jesuit observatories, sometimes acting as an intermediate for their purchase and

[10]*Que l'intérêt de l'institut optique ancien Fraunhofer puisse excuser mon arrogance!*

[11]*Il nous a accueilli d'une manière la plus bienveillante, nous lui avons fait ressortir les singulières perfectionnements de votre instrument, il a bien voulu prendre pratique lui-même de son maniement.*

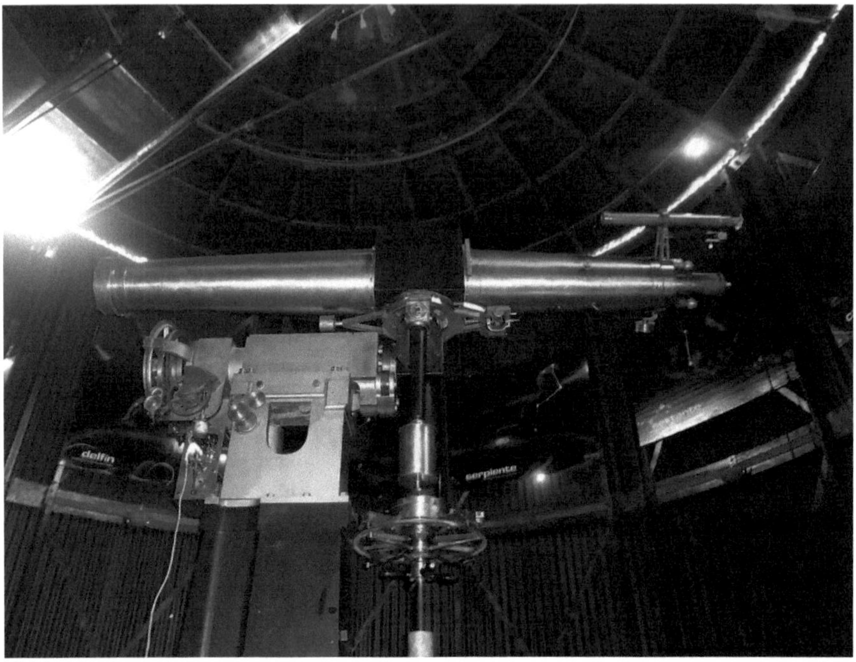

Fig. 4 Merz telescope still extant at Quito Observatory (Adam Kess—Opera propria, CC BY-SA 4.0, https://commons.wikimedia.org/w/index.php?curid=37096800)

negotiating sizes and prices. This is the case for the Merz telescopes that were installed in Calcutta, Manila, Quito (Fig. 4) and some European Jesuit observatories (Udìas 2003).

Sigmund Merz, on his side, often donated small instruments or accessories to Secchi, probably with the double intent of expressing his gratitude to the Jesuit astronomer and promoting his own production, as usually done by firms (Merz 1873a). The German instrument-maker was indeed very active in promoting his instruments, by participating in scientific events and astronomical congresses, like the one held in Vienna in 1869 (Merz 1869a).

3 Technical Questions

The main topic of the Merz-Secchi exchanges, however, concerned technical questions. From this point of view, their letters are a very precious source for historians, because they contain the descriptions of the solutions to several technical problems. In 1862, Secchi wrote to the Merzs that his objective-glass had been damaged by the atmospheric moisture; they explained to him how to remedy this:

> … you inform me that the objective-glass of your large telescope is damaged by atmospheric moisture. This problem is easy to be solved. You have just to clean it, namely to dry it with a soft canvas (some thin cloth), after unscrewing the ring and removing the lenses. The two lenses are separated by thin tin sheets. In order to replace the objective-glass in its frame, you need new tin sheets, which you will find here enclosed. They must be equally spaced, namely they must be placed at a distance of 120° from each other and pushed towards the center of the objective-glass. Be careful to fix the sheets with some glue (gummi arabicum would be the best) and to avoid the glue being too thick. It would be better to use any liquid one. Once the sheets are fixed, it is enough. The enclosed sheets are of the same thickness and carefully selected; it is also absolutely necessary that the glue coat does not alter their mutual thickness. Likewise, it is necessary to check that the two lenses (Crown & Flint) are put together in the same manner as they were originally.[12] (Merz Georg 1862b)

One year later, Secchi again met with similar problems:

> Now, regarding your 9-inch objective-glass, the stains that have appeared there again are undoubtedly produced by the rain. Had you been able to dry the objective-glass immediately after it rained, the moisture would not have left any stain. Regretting very much this unfortunate occurrence, I don't consider it to be an oxidation of the glass and I hope it will be remediable. You now have to try to make these stains disappear by treating them with saliva. If this is not enough, try saliva mixed with gypsum. I shall be honoured to send you some gypsum suitable for this purpose, together with the new eye-piece in question. We ourselves use it often, but it is a very delicate operation. In order to treat them as they need to be, it is best to use your fingers to scrub the stains. Having finished this manipulation, you have to take some good wine spirit and a very soft cloth to wash and clean the objective-glass completely. I hope that this is successful, otherwise the polishing of the objective-glass anew would require much time and struggle.[13] (Merz 1863b)

---

[12]*… vous communiquez que l'objectif de votre grande lunette soit gâté par l'humidité de l'atmosphère. Ce mal est facile à réparer. Vous n'avez que le nettoyer, c'est-à-dire que le faire essuyer par de toile molle (du linge fin). Après avoir desserré les vis qui tiennent l'anneau et prendre dehors les lentilles. Le deux lentilles se touchent par de feuilles minces en tain. Pour mettre l'objectif de nouveau dans la monture il faut des nouvelles feuilles en tain, les voilà y incluses. Elle doivent [être] mises tout à fait également c'est-à-dire mises en distance de 120° et poussées l'une aussi loin que l'autre vers le centre de l'objectif. Veuillez observer encore que cette feuilles seront fixées par de colle (gummi arabicum serait le mieux) et que la colle ne doit pas être trop épaisse. Il sera mieux de la prendre trop liquide. Si les feuilles tiennent alors il est assez. Les feuilles envoyées sont de la même épaisseur et scrupuleusement choisies, il est aussi bien nécessaire que la couche de colle ne fusse pas l'une plus épaisse que l'autre. De même il faut bien regarder que les deux lentilles (Crown and Flint) vont ensemble tellement mises comme elles sont sorties de chez nous.* On that occasion, Georg Merz wrote to Secchi that Sigmund would have been happy to do it by himself, during a possible visit to Rome and Italy.

[13]*En ce qui concerne maintenant votre objectif de 9 pouces, les taches y remarquées de nouveau sont sans doute le produit de la pluie. Si vous aviez pu essuyer l'objectif dans le premier moment après la pluie l'eau n'aurait pas laissé des taches. Beaucoup regrettant cette mauvaise chose je ne la tiens pas donc pour une oxydation du verre et j'espère qu'elle sera remédiable. Il faut maintenant tenter que cettes taches disparaissent en les travaillant par salive. Si cela ne suffit pas, par salive avec de la craie. J'aurais l'honneur de vous envoyer de la craie propre à ce but avec le nouvel oculaire en question. Nous-mêmes, nous en faisons usage plus souvent, mais c'est toujours une opération très délicate. Pour manipuler, comme il faut, vous vous servirez le mieux du doigt en y frottant les taches. La manipulation ainsi achevée il faut prendre du bon esprit de vin et du*

In November 1865, Merz constructed a spectrometer for Secchi (see below) and Sigmund exalted the quality of the glass prisms, though their colour was a bit dark:

> This is why we have chosen this glass for the spectrometer. The dispersive power is almost double the ordinary Flint. Consequently, the spectral lines (called Fraunhofer lines) became sharper or well multiplied. The solar spectrum actually appears to be very interesting. This glass is the result of our invention. I don't know of any other glass having such a high dispersive power. Here are the exponents of the refraction ratio: Cn = 1.7389, En = 1.7621, Dn = 1.7513, Fn = 1.7723.
>
> What a shame that this glass is not white too. I would not succeed, in spite of the six attempts which I have made. However, the superiority of the spectral analysis is recognized everywhere, and I hope the same is also true for your observations. For the current instrument, we cannot use another glass, because the incident and the sorting light beams must form an angle of 120° between them, while in [the case of] ordinary Flint, under the same ratio, they form only a 90° angle. In case you absolutely must have very white Flint prisms, it is necessary to return the whole instrument. Then I will propose reshaping the device so that the prisms can be changed. This could be done easily through the intermediate piece joined to the current device.[14] (Merz 1865b)

The problem of the coloured glass arose again in 1874, when Secchi acquired a new direct-vision prism:

> About the direct-vision prism, I regret very much that the glass is a little dark, but the images are excellent and, for the moment, I cannot offer you a better prism of this size.[15] (Merz 1874b)

Sometimes, the Merzs explain the use of certain accessories and/or the reason for certain adjustments:

(Footnote 13 continued)

linge très mou pour laver et tout à fait nettoyer l'objectif. J'espère que vous y réussirez car s'il fallait de polir l'objectif de nouveau il couterait du temps et de la peine.

[14] *Pour avoir choisi ce verre pour le spectromètre voilà la raison. Le pouvoir dispersif en est presque le double que dans le Flint commun. Par conséquent aussi les raies du spectre (dites lignes de Fraunhofer) deviennent plus distinctes ou multipliées. Le spectre solaire en est en effet fort intéressant. Le dit verre est le produit de mon invention. Je ne connais aucun verre d'un aussi grand pouvoir dispersif. En voici les exposant des rapport de réfraction: Cn = 1.7389, En = 1.7621, Dn = 1.7513, Fn = 1.7723. Bien dommage que ce verre n'est pas aussi blanc. Je n'y réussirais pas malgré les six épreuves que j'ai fait. Cependant la supériorité pour l'analyse spectrale en est reconnue partout et j'espère qu'il s'éprouvera aussi dans vos observations.*

Pour l'instrument actuel nous ne pourrions pas faire usage d'un autre verre, parce que le rayon émergeant y doit former un angle de 120° avec le rayon incident, pendant que le Flint commun sous le même rapport ne forme que 90°.

En cas que vous voudriez absolument avoir des prismes d'un Flint très blanc il fallait avoir tout l'instrument renvoyé. Alors je proposerais encore de former l'appareil ainsi qu'on pourrait changer les prismes. Cela se ferait facilement par la pièce intermédiaire qui se trouve déjà à l'appareil actuel.

[15] *Quant au prisme à vision directe je regrette beaucoup que les verres sont un peu sombres, mais les images en sont excellentes et pour le moment je ne peux pas vous offrir un meilleur prisme de cettes (sic) dimensions.*

> The reflexing prism of the telescope must be placed in front of the eye-piece. Its positioning between the lenses would have required a prism of such a size that the rod for adjusting the eye-piece would have become much too short, but it is necessary to have it being a bit longer, to be able to turn it comfortably in its position.[16] (Merz 1874b)

And later, about a direct-vision spectroscope: "*The prisms are glued in a new manner which I consider to be the best*"[17] (Merz 1876a). Sigmund Merz's effort for improving the instruments and satisfying the requirements of his renowned commissioner is quite clear.

The use of a cylindrical lens for spectroscopic observations is often dealt with in the Secchi-Merz correspondence. It was a very controversial topic, because it was difficult for early astrophysicists to find a proper combination of these lenses with the slit and the eyepiece.[18] In 1867, Secchi asked Merz to make a special eye-piece, including a cylindrical lens for his spectral apparatus; Sigmund replied:

> Mr. Fraunhofer used a simple cylindrical lens, but it has never been combined with an eye-piece (Ramsden system) until now. Some years ago, I spoke about this with Mr. Lamont, who also proposed the use of a cylindrical lens to observe the spectra. I am presently ready to build a machine for the exact execution of cylindrical surfaces.[19] (Merz 1867d)

Later, Secchi tried to replace the slit with a cylindrical lens, but Sigmund expressed his perplexity about this expedient:

> I believe now that slit spectroscopes are still necessary, being that they are the only ones which can be used to observe the spectra of the Sun, the Moon and extended nebulae. Since the light beams passing through the direct-vision prism change their focal plan, I well understand why you use the concave half-lens. [...] But [...] the slit and the lens can be used by choice, because the cylindrical lens produces the image of the star as a line exactly in the slit plan. The slit, when opened wide, does not alter the star's image, while the narrowed slit collects just a portion of the light coming from nebulae, and consequently forms a bright line which is impossible to be produced by the cylindrical lens alone, giving just a large band [in the case] of nebulae. Why would you put the axis of the cylindrical lens perpendicular to the slit? You will reply: in order to get light and width in the spectrum. This is surprising for me if I think that most of the light cannot pass through the slit.[20] (Merz 1868c)

---

[16] *Le prisme à réflexion de la lunette se mets devant l'oculaire. L'arrangement entre les lentilles aurait demandé un prisme d'une telle dimension que le tirage pour ajuster l'oculaire serait devenu beaucoup trop court mais il faut l'avoir un peu large pour le tourner commodément en position.*

[17] *Les prismes sont collés d'une nouvelle manière dont je pense qu'elle soit la meilleure.*

[18] Merz delivered a cylindrical lens to Secchi in 1863 and another one in 1866 (see Wolfschmidt and Kost, in this volume). On the use of cylindrical lenses in spectroscopic apparatus, see Hearnshaw 2009.

[19] *Des simples loupes cylindriques s'ont servi déjà M. Fraunhofer, mais composés pour un oculaire (système Ramsden) on ne les avait pas employés jusqu'ici. Il y a quelques ans que j'en avait parlé à M. Lamont, qui proposait pour les observations des spectres encore un objectif cylindrique. A' présent j'y suis [de prêt?] à faire construire une machine pour pouvoir exécuter exactement des surfaces cylindriques.*

[20] *Je crois maintenant que le spectroscope à fente soit encore nécessaire, lui seul pouvant servir aux observations du spectre du soleil, de la lune et des nébuleuses de grand diamètre. Vu que les*

Merz was often ready to suggest the cheaper solution for Secchi's requirements:

> About the spectroscope [...] In order to compare the direct image with the spectral lines, I would always prefer the concave lens, as it will be cheaper, while the prism in question requires much work.[21] (Merz 1868d)

A recurrent technical problem was the irregular speed of the clockdrive mechanism of the Merz refractor at the Collegio Romano Observatory. It is quite known that this was probably the main weak point of the large Merz refractors, because the mechanism often turned out to be unreliable, especially in the first generation of these instruments, built in the 1850s. Secchi pointed out this defect several times; Merz, however, was aware of the problem: "*What you say about the clock of your equatorial [telescope] is right, indeed. We have since abandoned this system. I shall not forget to send you a draft of the new system with a conical pendulum in another letter and I don't doubt that it will still be possible to modify the system of your instrument in this way*[22] (Merz 1868d). In 1868, Sigmund sent Secchi the drawing and description of the new conical pendulum-driving mechanism which he had adopted for the Brera and Naples refractors and which assured regular functioning (Merz 1868e).

4 Not Only Telescopes: Spectroscopic Equipment by Merz

The most important partnership between Secchi and the Merzs concerned the development of a spectroscopic apparatus for analyzing starlight. Secchi is recognized as one of the pioneers in the spectral classification of stars (Fig. 5) (Hearnshaw 1991) and was able to obtain such important results thanks to his close cooperation with Merz, from the beginning of his studies.

(Footnote 20 continued)

rayons ayant passés le prisme à vision directe changent leur foyer je comprends bien que vous employez la ½ lentille concave [...] Mais [...] on peut faire usage à volonté de la fente ou de la lentille, vu que la lentille cylindrique produit l'image de l'étoile réduite en ligne tout à fait dans le plan de la fente. La fente grandement ouverte n'altère pas l'image d'une étoile, la fente diminuée laisse seulement passer une portion de la lumière des nébuleuses, formant de cette manière une ligne lumineuse impossible à faire par la lentille cylindrique seule qui ne peut que donner une bande large des nébuleuses. A quel but mettez-vous l'axe de la lentille cylindrique perpendiculaire à la fente? Vous répondrez: pour avoir lumière et largeur dans le spectre. Cela m'étonne en pensant que la plus grande partie de la lumière ne puisse pas passer par la fente.

[21]*Quant au spectroscope [...] Pour comparer l'image directe avec les lignes du spectre je préférais toujours la lentille concave vu qu'elle sera meilleur marché, car le prisme en question fait beaucoup de travail.*

[22]*C'est-ce que vous dites de l'horloge de votre équatorial a bien raison. Il y a déjà long temps que nous avons quittés ce système. Je ne manquerai pas de vous donner l'esquisse du nouveau système à pendule conique dans une prochaine lettre et je ne doute pas qu'on pourrait encore changer le système de votre instrument de cette manière.*

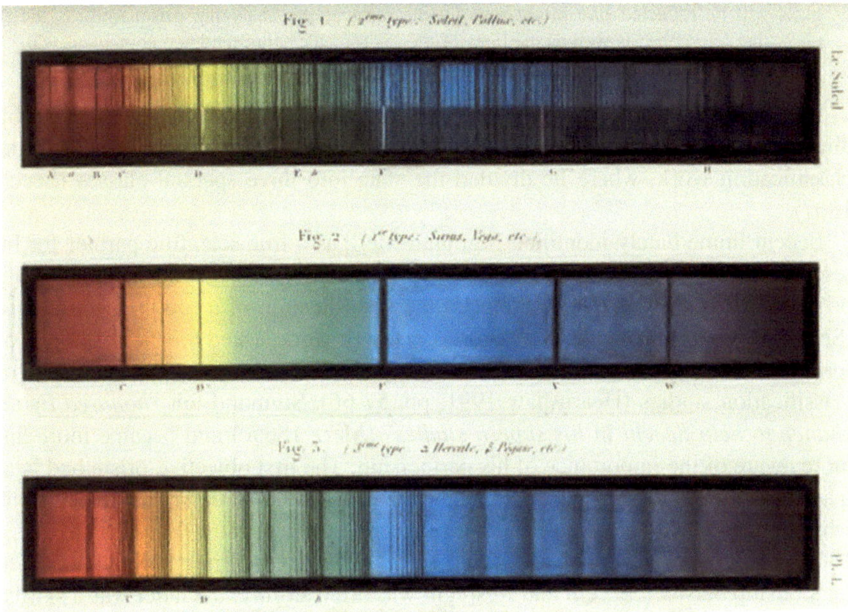

Fig. 5 Secchi's first spectral classification of stars (from: Secchi 1875–1877)

After some experiences in stellar spectroscopy carried out in November 1862 with the French astronomer Jules C. Janssen (1824–1907) at the Collegio Romano Observatory, in 1863, Secchi started his own research program on the spectral classification of stars. He needed a good spectroscope for stellar observations and addressed his request to the German firm Ertel and Sons,[23] which usually built the mechanical pieces for Merz optical instruments. Ertel forwarded Secchi's request to the Merzs, who communicated their ability to construct the required instrument in 4–6 weeks, at a price of 100–150 francs:

> From Mr. Ertel, we learn that you wish to provide your large refractor with a device for observing the star spectra, similar to the device whose construction is described by Mr. Littrow in the [journal] Cosmos [edited] by Mr. Moigno. [...] The device, screwed in by replacing the eye-piece, is equipped with a cylindrical lens, whose focal plan is equal to its distance from the focal plan of the objective-glass of the large refractor, thus reproducing the star as a bright line. In front of the glass, a Flint 60° analyzing prism will be placed, and in front of this prism, a small telescope will be adjusted for parallel rays, and beside this telescope, a second tube will be placed, having a micrometer whose golden lines ruled on

[23]Ertel facilitated the exchanges between Secchi and Merz; his role, however, deserves to be clarified more precisely.

> glass will be reflected like bright rays by the facet of the analyzing prism towards the observing telescope, as well as the lines of the micrometer being rendered coincident with the spectrum received by the same apparatus.[24] (Merz 1863a)

In November 1865, Merz delivered a spectrometer to Secchi with a double direct-vision prism system, which was used by the Jesuit for his first spectral classification work, where he divided the stars into three spectral classes (Secchi 1867).

Secchi immediately identified Sigmund Merz as a true scientific partner for his research program. In 1869, he sent him (Merz 1869b) his second memoir on the prismatic spectra, where he announced the identification of a fourth spectral class (Secchi 1869) and, in order to extend his research work, asked the German optician for an objective-prism, which became the main instrument for Secchi's spectral classification studies (Hearnshaw 1991, pp. 57-66). Sigmund felt "*honored by the chance to help Secchi in his superb studies*" (Merz 1869c) and became more and more aware of the importance of his partnership. The first objective-prism had been made by Ertel for Lorenzo Respighi (1824–1889), director of the Capitol Observatory in Rome, another Italian pioneer in astrophysics, who designed it in 1868 for his spectral studies, taking up an original idea by Fraunhofer (1817). The relationship between Secchi and Respighi was often stormy: the latter was a skittish personality and suffered the scientific rivalry with the brilliant Jesuit, who enjoyed wide popularity. Respighi probably kept the details of the instrument undisclosed, and Secchi turned to Merz: "*Prof. Respighi told me wonderfully about the prism you have sent to him*"[25] (Merz 1870, p. 164), he wrote, asking for further information, as he intended to buy one. Sigmund replied:

> Regarding the prism of Prof. Respighi, it is 130 mm -about 4,5 inch- large and its cost was 210 f[iorins] = 450 fr[ancs]. As a sign of gratitude to you, dear Father, and in the interest of these studies which seem to me to be reinvigorating the method of spectral observation of the late Mr. Fraunhofer, I offer you also a 6-inch prism, in case you prefer to have it cut round, at the same price of 450 francs.
>
> Indeed, for a larger prism, it would be necessary to raise the price, since an 8-inch prism requires a piece of good glass having the same weight as a 12-inch objective-glass! The refraction angle [is] 15 degrees and the surfaces' shape [is] round.

---

[24]*C'est par Mr. Ertel nous apprenons que vous souhaitez pour votre grand réfracteur un appareil pour produire les spectres des étoiles, semblable à cet appareil dont la construction Mr. Littrow a donné dans le Cosmos de M. Moigno. [...] L'appareil vissé au lieu de l'oculaire est muni d'un verre cylindrique,dont le focus est égale à sa distance du focus du grand objectif du réfracteur, produisant alors une ligne lumineuse de l'étoile. Devant le verre sera mis le prisme analysateur (en Flint) de 60° er devant celui-ci une petite lunette accommodée pour des rayons parallèles et coté de cette lunette se trouvera un second tuyau avec un micromètre dont les raies tracées en or sur verre seront réfléchies comme raies lumineuses par la surface du prisme analysateur qui porte vers la lunette d'observation ainsi que les raies du micromètre peuvent être amenées en coïncidence avec le spectre reçu par l'appareil même.* The Merz Company guaranteed the delivery of the instrument in 4–6 weeks, at a cost of 100–150 florins.

[25]*Le Prof. Resphigi* [sic] *me dit des merveilles du prisme que vous lui avez envoyé.*

> If I ask you for 1000 francs for an 8-inch diameter prism, I am sure you will find it reasonable, because the above-mentioned amount is really just the price of the glass in the rough.[26] (Merz 1869c)

Secchi asked for an objective-prism larger than Respighi's and having special requirements; for this reason, he started negotiations with Sigmund, who wrote to him:

> I have the drawing for making a ring with three clamping screws, like Mr. Ertel has done for the Capitol [Observatory] equatorial as a mounting for the prism. Regarding the other support, do you think, Sir, that it is absolutely necessary that the prism be able to rotate completely on an axis which is perpendicular to the optical axis of the telescope? Suppose that the prism is placed in front of the objective-glass so that the emerging rays could be directed to the optical axis of the telescope and that the prism is provided with a small correction so that the surfaces of the spectral prism, namely the axis of this prism, could be parallel to the objective plan.
>
> What would you say, Sir, if I put the spectral prism out of the center of the objective-glass, in order to have more room for the other achromatique prism, which you propose for the direct rays, considering that just one inch of space seems to me to be too little?[27] (Merz 1869d)

A few months later, Sigmund wrote to Secchi that the construction of "*the large 6-inch prism was well advanced*" and that he was encountering greater difficulty than he thought he would[28] (Merz 1869e). Once ready, the objective-prism was sent

---

[26]*Quant au prisme de Mr. le Prof. Respighi il a une largeur de 130 mm à peu près 4½ pouces et il a couté 210 f[lorins] = fr[ancs] 450.*

Dans les sentiments d'obligeance envers vous mon cher Père comme dans l'intérêt de la chose même qui me semble recommencer la méthode d'observation spectrale du feu Mr. Fraunhofer je vous offre même un prisme de 6 pouces en cas qu'il vous convienne de l'avoir facetté rond pour le même prix de 450 francs.

Mais pour un prisme plus large il faudrait bien hausser le prix, vu qu'un prisme de 8 pouces exige déjà un morceau de bon verre d'un poids aussi grand comme un objectif de 12 pouces! L'angle de réfraction y compris de 15 degrés et les surfaces en forme ronde.

Si je vous demands 1000 francs pour un prisme de 8 pouces en diamètre je ne doute pas que vous le trouverez raisonable, car le dit montant est vraiment le prix du verre brut.

[27]*J'ai le dessein de faire un anneau à trois vis de pression, comme Mr. Ertel l'a fait pour l'équatorial du Capitol comme support du prisme.*

Mais quand à l'autre armure pensez-vous, Monsieur, qu'il soit absolument nécessaire qu'on puisse tourner le prisme spectral pleinement autour d'un axe normal sur l'axe optique du télescope?

Supposé que le prisme soit mis devant l'objectif tel que les rayons émergentes puissent suivre la direction de l'axe optique du télescope et que le prisme soit pourvu d'une petite correction pour que les surfaces du prisme spectral, c'est-à-dire l'axe de ce prisme puisse être mis parallèlement au plan de l'objectif.

Et que dites-vous, Monsieur, si je mettais le prisme spectral hors du centre de l'objectif pour avoir plus de place pour l'autre prisme achromatique, que vous proposez pour les rayons directs, la largeur d'une pouce seulement m'en semblant trop petite?

[28]*Le grand prisme de 6 pouces est aussi bien avancé. Il me fait plus de peine que j'avais pensé.*

to Secchi, who immediately tested it, obtaining excellent results:[29] "*The aperture of the refractor is thus reduced by more than half of its surface [...] but, in spite of this, the light is so intense that it greatly surpasses that obtained by interposing direct-vision prisms near the eye-piece. The dispersion is so remarkable that it is at least six times larger than the one I have obtained with the most powerful spectroscopic eye-piece and even with the combined slit-less spectroscope, with a cylindrical lens, which I initially used*"[30] (Merz 1870, p. 165).

The price of the prism being very high, Sigmund gave Secchi the opportunity to pay for it in installments:

> I am delighted to learn that you are happy with your large spectral prism. At the same time, I must say that I would be very happy if you could pay what is possible for you, by dividing the amount into two or three payments at your convenience.[31] (Merz 1869f)

Sigmund Merz soon became an expert in the construction of spectroscopes; in 1868, he furnished one of his spectroscopes for the large refractor at the Moscow Observatory (Merz 1868a) and improved Hofmann's direct-vision spectroscopes[32] (having a 6° dispersion) up to 8°–12° dispersion (Merz 1868b). In a catalogue dated July 1, 1872, and signed by Sigmund Merz (1872b), nine spectroscopic apparatuses are listed (entries nos. 78–86). It should be remarked that the first (and most expensive) entry is the objective-prism (Fig. 6), described as "*Apparat der Sternwarte zu Rom*" ("*instrument of the Rome Observatory*"), while the fourth entry is Secchi's star spectroscope, explicitly presented as a "*simple star spectroscope by Fr. Secchi with cylindrical lens and direct-vision prism*" ("*Stern Spektralapparat, einfacher nach Pat. Secchi mit Cylinderlinse und Prisma à vision directe*"): once more, the collaboration with Secchi is well "advertised" by Merz. With the same purpose, Merz explicitly mentioned Secchi's advice about the objective-prism in some "promotional" news published in a technical journal (Merz 1870a).

Secchi's spectral classification studies mostly covered the years 1867–1872, but, even later, the Jesuit did not stop improving his observations and requiring other spectroscopic instruments; in 1874, Merz replied:

[29]It should be mentioned that Respighi disputed Secchi over the priority of the use of the objective prism; actually, Respighi was the first to use this kind of device, but Secchi applied it methodically in his investigations.

[30]*L'ouverture du réfracteur reste àinsi réduite de plus de la moitié de sa surface [...] mais, malgré cela, la lumière est si intense qu'elle surpasse beaucoup celle qu'on obtient avec l'interposition des prismes à vision directe près de l'oculaire. La dispersion est si considerable qu'elle surpasse, de six fois au moins, celle que j'ai obtenue avec l'oculaire spectroscopique plus puissant, et même avec le spectroscope composé sans fente, à lentille cylindrique, que j'ai employé primitivement.*

[31]*Que vous êtes content du grand prisme spectral j'apprends avec plaisir. En même temps je peux vous dire que je suis bien content si vous payerez pour le montant tout ce qu'il vous est possible en partageant le payement en deux ou trois tout à votre aise.*

[32]On direct-vision spectroscopes, see: Warner 1993.

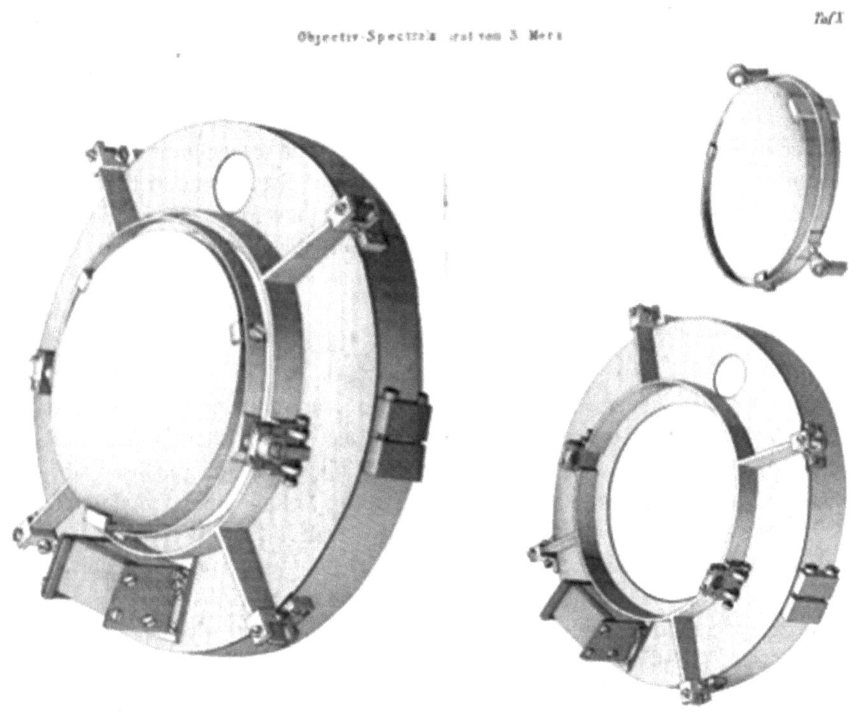

Fig. 6 Merz's illustration of the objective-prism (from: Merz 1870)

> Regarding the price of a rectangular prism to be placed in front of the objective-glass of a 4-inch telescope, I believe I will be able to construct it for 1500 francs. The direct-vision prism which you request will be made as soon as possible. Unfortunately, the stock of good, very heavy flint [glass] is depleted at the moment. [...] you [also] need an eye-piece for your spectroscope ...[33] (Merz 1874a)

After 1873, the financial resources of the Collegio Romano Observatory were drastically reduced.[34] In order to economize, Secchi sometimes asked Merz to reshape or recycle some prisms; Merz, however, often did not succeed in this operation:

---

[33]*A ce qui concerne maintenant le prix d'un prisme rectangulaire placé devant l'objectif d'une lunette de 4 pouces, je crois de pouvoir faire un tel à frs. 1500.*

Le prisme à vision directe, que vous me demandez sera commencé aussitôt. Malheureusement la provision de bon flint très lourd n'est pas trop grande dans ce moment. [...] vous avez [aussi] besoin d['un] oculaire pour votre spectroscope ...

[34]After the annexation of Rome to the Kingdom of Italy in 1870, all ecclesiastic properties were gradually seized by the Italian Government. The Collegio Romano Observatory was not confiscated, in consideration for Secchi, but neither the Papal State nor the Italian Government intended to invest in an observatory whose destiny was unclear.

> ... regarding this prism, I regret very much not being able to better reshape it. The glass of the prism has been used for a long time. The prism also shows a small crack and it could get broken if it is warmed up further[35] (Merz 1876b)

The last letters from Sigmund Merz to Secchi, which are kept at the Gregorian University, date back to 1876. Financial straits, as well as health problems, limited Secchi's scientific activity in the last years of his life. Secchi died in 1878, when he was just 59 years old, but the way to the "new astronomy" and its consolidation as a scientific discipline (Chinnici 2008), had definitely been paved.

5 What Is Extant?

In addition to the letters, which deserve a more careful examination,[36] other historical materials, related to the Merz-Secchi connection, are extant, mostly in Rome. Unfortunately, the beautiful refractor of the Collegio Romano Observatory got destroyed (see Altamore, Poppi and Maffeo in this volume), but a small collection of Secchi's spectroscopic instruments, built by Merz, is still preserved at the Rome Astronomical Observatory. In addition, five rare photographs of those apparatuses are kept at the Science Museum in London[37]. The photographs date back to 1876, when a special loan exhibition of scientific apparatuses was set up at South Kensington.[38] On that occasion, Secchi sent some photographs of his spectroscopic equipment: today, these documents are an important historical source, as they show the instruments as they originally appeared, including some lost spectroscopes. Below, a description of what is extant of the Merz instruments that were used by Secchi.

5.1 Star Spectroscope

This instrument (Fig. 7) is very similar to the spectroscope described by Secchi in 1867 (Fig. 8) in his first memoir on the spectral classification of stars (Secchi 1867,

[35]*... à ce qui concerne le prisme en question, il me faut bien regretter de ne pouvoir pas le renouveller* (sic) *mieux. Le verre, dont le prisme est composé, est long temps consommé. Le prisme a aussi une petite fêlure qu'il pourrait bien casser si l'on voulait échauffer plus fort.*

[36]Thirty letters by Ertel to Secchi are kept at APUG; they could contain additional information about Secchi-Merz connections (see note 23).

[37]The five photographs bear the inventory numbers 1876–683, 1876–684, 1876–685, 1876–686, 1876–687; the author is the Roman photographer Giuseppe Della Valle, as can be argued by a dry stamp on their verso. Two of these photographs are here reproduced in Fig. 11 and 13; the others represent a reflection spectroscope (photograph No. 1876–687), probably made by Secretan and illustrated in *le Soleil* (Secchi 1875–1877, I, pp. 227–229), and a spectroscope with a diffraction grating by Lewis M. Rutherfurd (1816–1892) which is attached to the focus of the Merz refractor (photographs No. 1876–685 and 1876–686). Both spectroscopes are no longer extant. On diffraction gratings, used instead of prisms for light dispersion, see Hearnshaw 2009, pp. 17–24.

[38]On the history of the Science Museum and its collections, see Morris 2010.

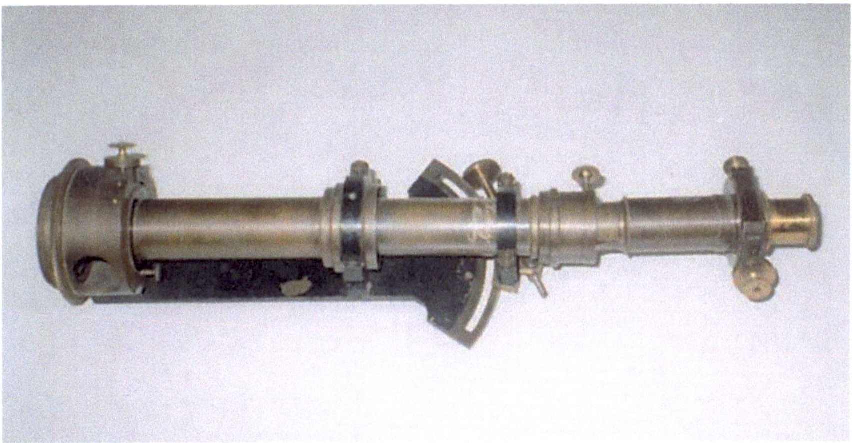

Fig. 7 Secchi's star spectroscope by Merz (courtesy of INAF-OAR)

p. 72). It is a 5-prism direct vision signed *G. and S. Merz in München*. It appears in the aforementioned Merz catalogue of 1872 (entry no. 81), in which it is credited to Secchi, and it is also illustrated in Secchi's treatise *Le Soleil* (Fig. 9). The Jesuit tested this instrument by interchanging several direct-vision prisms and replaced the spider wires of the micrometer—which were too difficult to be seen while observing faint stars—with two steel tips (Fig. 10), overlapping the spectral line to be measured without covering it completely (ibid., p. 73). Secchi intended to use it as a slit-less spectroscope, only with a cylindrical lens. For this reason, he appreciated—and mentioned in his memoir—that "*Mr. Merz has proposed a very practical device for widening the slit, without removing the supporting diaphragm, so much as to render it unnecessary, as if it were not even there*"[39] (ibid.). In spite of some successful observations, Secchi was unhappy with the performance of the instrument and preferred to use it in a simplified version - just placing the prism between the cylindrical lens and the eye-piece (ibid., p. 72). Today, the instrument is in a passable condition, though some original parts appear to be missing or badly assembled, and is kept in the historical instruments collection of the Rome Astronomical Observatory, at Monte Porzio Catone, near Rome.

5.2 *Photograph of Secchi's Solar Spectroscope*

This spectroscope is no longer extant in the Rome Observatory collections (Calisi 1991, pp. 98–99; 103): there is only one existing coeval photograph (Fig. 11) that

[39] *Il Sig. Merz ha proposto un meccanismo di fessura assai comodo, e che senza levare dal posto il diaframma che la porta, permete di allargarla tanto che si rende inutile e come se non vi fosse.*

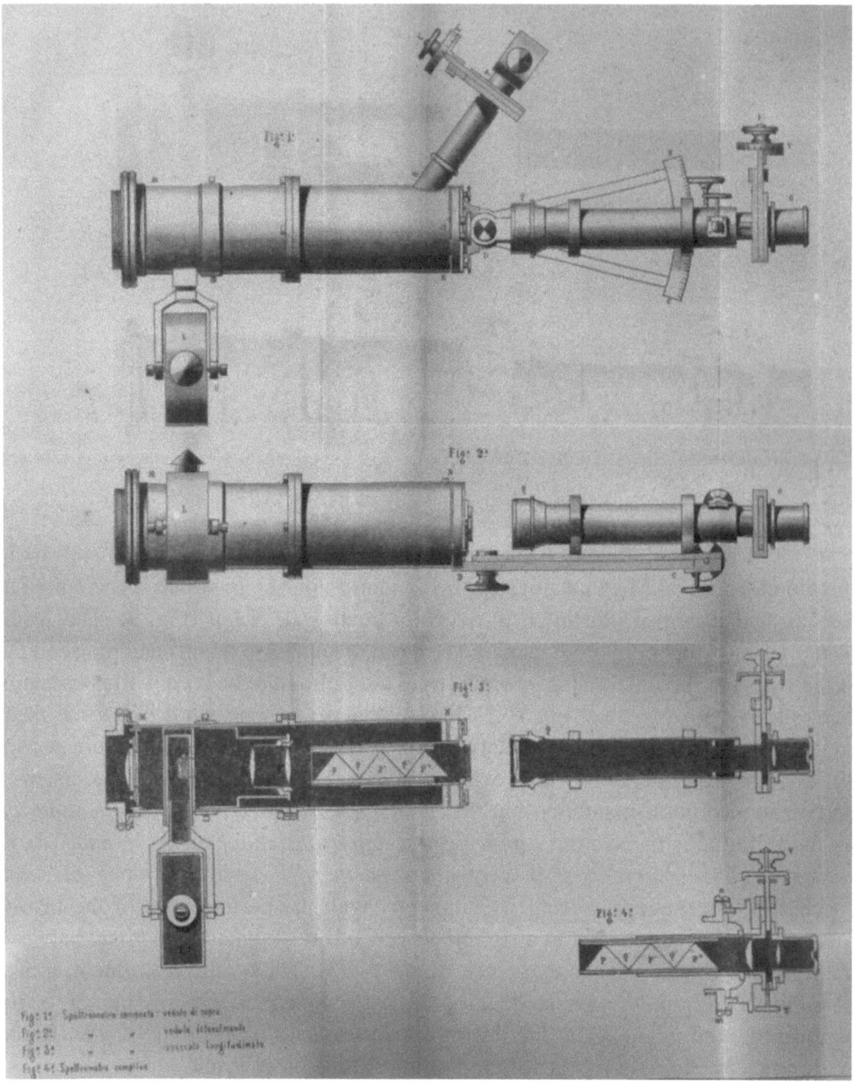

Fig. 8 The spectroscope by Merz as described in Secchi's first memoir on the spectral classification of stars (from: Secchi 1867)

shows it, at the Science Museum in London;[40] consequently, it is a unique document, regarding this lost instrument.

The spectroscope in the photograph is a combination of a direct-vision prism with three angular prisms. This device was preferred by Secchi for observing solar

[40]The photograph is inventoried with No. 1876–683.

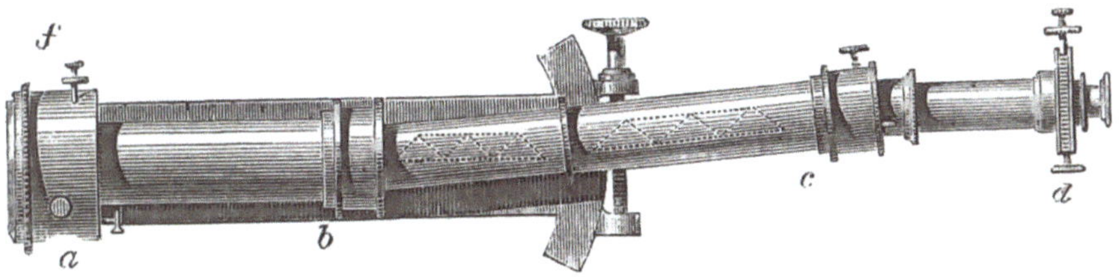

Fig. 9 The star spectroscope, illustrated in Secchi's treatise *Le Soleil* (from: Secchi 1875–1877, I, p. 231)

Fig. 10 The two steel tips placed in the eye-piece in order to help measuring the position of the spectral line (photograph by Aldo Altamore, courtesy of INAF-OAR)

prominences, as he stated in the second edition of his book *Le Soleil*, in which a picture of this instrument also appears (Secchi 1875–1877, I, p. 229) (Fig. 12). The direct-vision prism, made of 5 prisms, is not visible, as it is inserted in the end part of the tube on the left. This system, probably dating back to 1863, is an improvement on a similar combination with two angular prisms, well-described and illustrated in the first edition of *Le Soleil* (Secchi 1870, p, 200) as well as in Secchi's treatise on stars (Secchi 1895, I, p. 77). It had the advantage of easily increasing or diminishing the number of prisms, thus changing the setting of the

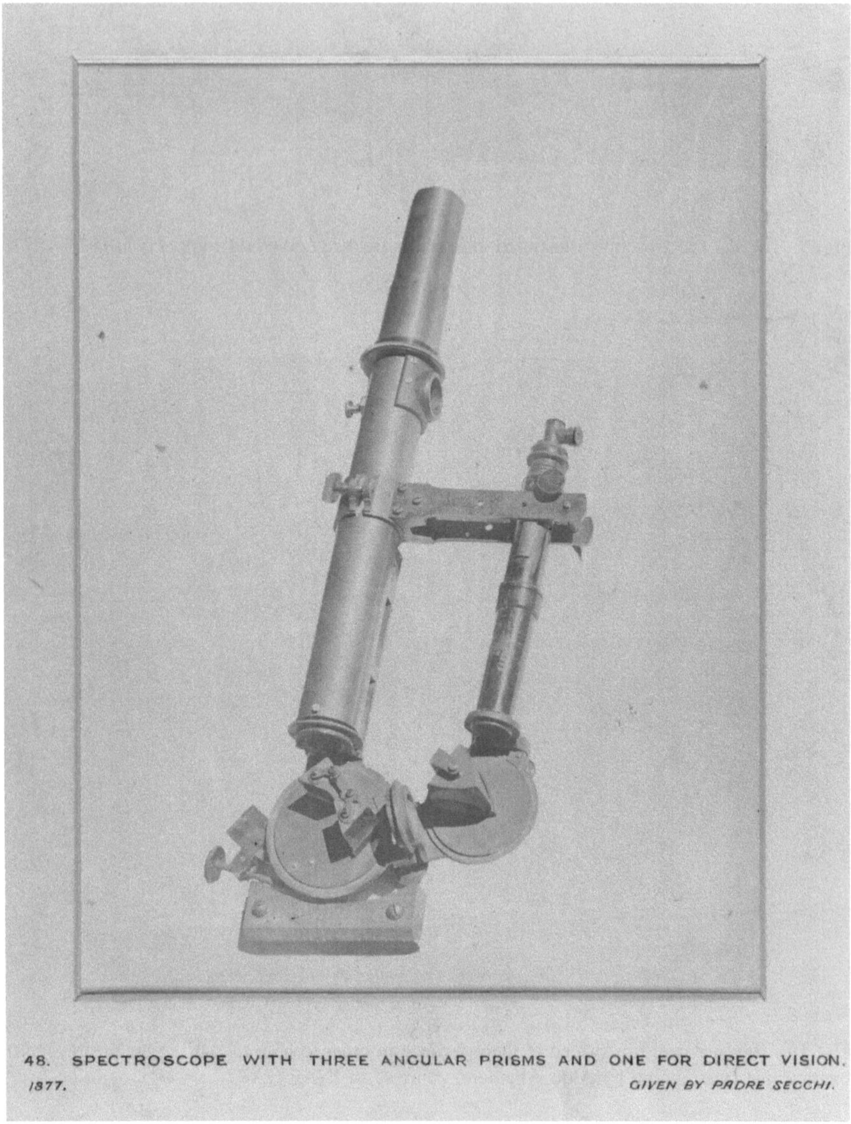

Fig. 11 Rare photograph of Secchi's solar spectroscope (courtesy of Science Museum, London)

Fig. 12 Illustration of Secchi's solar spectroscope (from: Secchi 1875–1877, I, p. 229)

instrument, so as to be able to observe different celestial objects.[41] In effect, in 1875, Secchi wrote that this is the spectroscope "*we usually make use of, whose power can be modified by adjusting the number of prisms from one to five.*"[42] (Secchi 1875–1877, I, p. 229). It is highly probable that this is the "*powerful spectroscope by Merz*" ("*un puissant spectroscope de Merz*") mentioned by Secchi in his works (Secchi 1875–1877, II, p. 84).

5.3 Objective Prism

In 1877, Secchi stated that "*the most proper device for these investigations* [stellar spectroscopy] *is certainly the angular prism, with a small refracting angle, put in front of the object-glass*"[43] (Secchi 1895, p. 447). The objective prism,[44] manufactured by Merz, is made of very pure flint glass and has an apex angle of about 12° and a diameter of 6 inches (=16 cm). It was supported in a suitable armature to be placed in front of the 9-inch object glass of the Merz equatorial telescope of the Collegio Romano Observatory. The prism could rotate though two *tourillons*, to obtain the minimum dispersion angle. In a photograph from 1876 (Fig. 13),[45] the instrument appears assembled, as first illustrated by Merz (Fig. 6) in 1870 (Merz 1870); pictures of the disassembled parts are also represented in Secchi's treatises (Fig. 14) (Secchi 1875–1877, II, pp. 448–449; 452–53; and Secchi 1895, pp. 80–83). In the armature, there was a small prism (visible in the photograph), which should have been used for receiving the direct image of the star, but the results were unsatisfactory, and Secchi decided to observe the direct image of the star, as usual, with a small guiding telescope (Secchi 1875–1877, II, p. 448). However, by attaching the objective prism to the large Merz refractor, he obtained too wide, and consequently faint, spectra. For this reason, he preferred to use the objective prism with his 8-inch aperture Cauchoix telescope, as "*this preserves light and gives more*

[41]Secchi stated in 1877: "*Direct-vision spectroscopes are more manageable but they extend the instrument a great deal and, in some cases, due to the sizes of the mobile domes of the observatories, they cannot be used. Moreover, in an angular prism spectroscope, it is easy to reduce the dispersion by removing a prism, and thus to adjust the instrument so as to observe fainter objects, which is impossible to do with direct-vision spectroscopes*" ("*Les spectroscopes à vision directe sont plus commode, mais ils allongent beaucoup l'instrument, et, dans certains cas, en raison des dimensions des dômes mobiles des observatoires, ils ne peuvent pas être employés. De plus, dans un spectroscope à prismes angulaires, on peut facilement enlever un prisme et diminuer la dispersion; on l'adapte ainsi à des objets plus faibles, ce qui ne peut se faire avec les spectroscopes à vision directe*") (Secchi 1875–1877, II, p. 76).

[42][Voici] *le spectroscope dont nous faisons habituellement usage, et dont on peut modifier la puissance en y adaptant à volonté un nombre de prismes variant de un à cinq.*

[43]*… l'appareil le plus propre à ces recherches est certainement le prisme angulaire d'un faible angle réfringent, que l'on place avant l'objectif ….*

[44]About the history of the objective prism spectroscopy, see Hearnshaw 2009, pp. 143–161.

[45]This is the photograph, bearing the inventory number No. 1876–684 at the Science Museum.

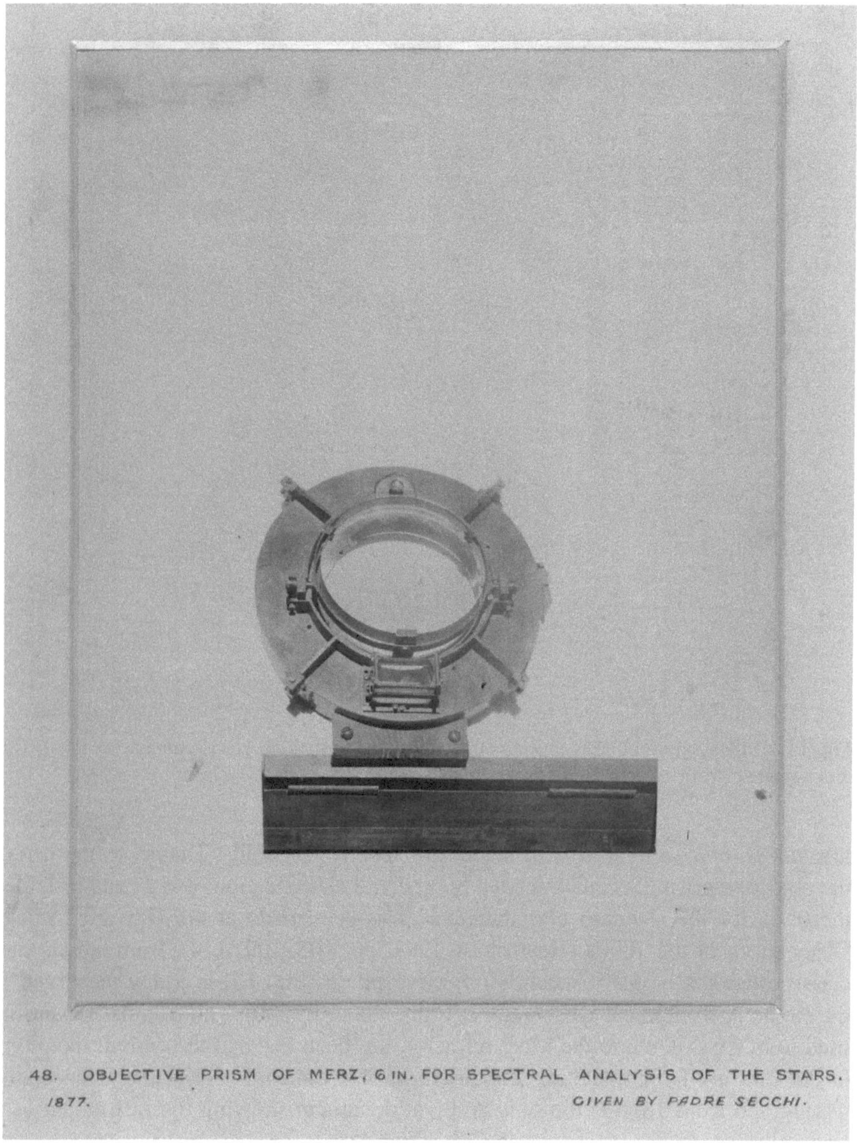

Fig. 13 Rare photograph of Secchi's objective prism by Merz (courtesy of Science Museum, London)

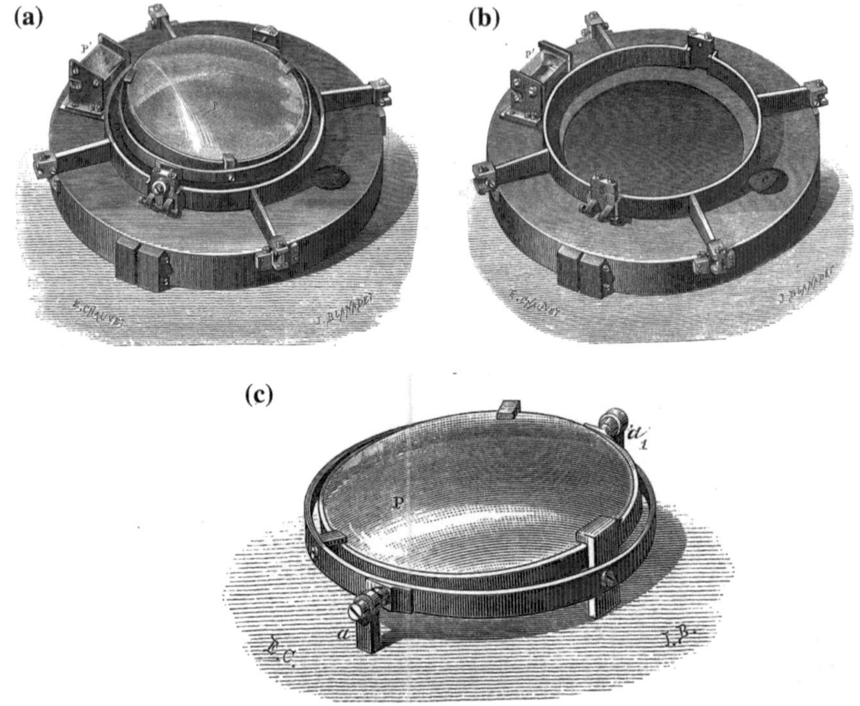

Fig. 14 **a** Illustration of the objective-prism; **b** the armature, disassembled; **c** the prism, disassembled (from: Secchi 1895, pp. 449; 452–53)

brightness, and you can obtain wonderful spectra."[46] (ibid.) Thanks to the use of the objective prism, Secchi extended his spectral classification of stars and provided the basis for the standard classifications that were made at the Harvard College Observatory in the 1890s (Hearnshaw 1991, pp. 108–109; 134), from which later classifications derived.[47] Secchi's objective prism (Fig. 15) is today preserved at the Rome Astronomical Observatory (Calisi 2009, p. 200). The original mounting, used to fit the prism into the Merz refractor, has been lost and the current mounting is the one used for attaching the prism to the Cauchoix telescope. Hence, the photograph from 1876 is the only archival document showing the instrument as it originally looked.

[46]*... on ne perd rien en lumière et l'on a plus de vivacité, de sorte que l'on peut avoir des spectres magnifiques.*

[47]For a historical survey of atlases of stellar spectra, see Hentschel 2002, p. 361.

Fig. 15 Secchi's objective prism today (photograph by Francesco Poppi, courtesy of INAF-OAR)

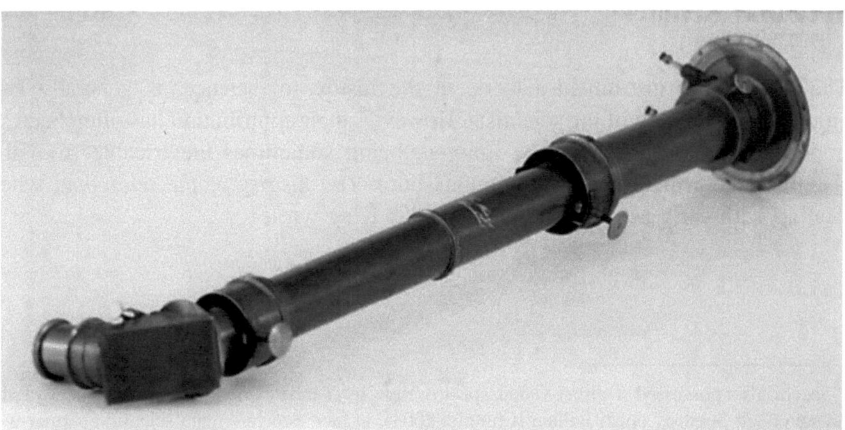

Fig. 16 Direct-vision spectroscope by Merz (courtesy of INAF-OAR)

5.4 Direct-Vision Spectroscope

It is possible that this (Fig. 16) is the spectroscope which is mentioned in some letters, dated from 1872 (Merz 1872a) and recorded in Merz's Account Books (see

Wolfschmidt and Kost in this volume); it is probably the universal spectroscope No. 80, described in the catalogue of 1872 (Merz 1872b; Merz 1870, p. 273), an instrument which could be used both as a simple and combined spectroscope.[48] This one, signed *G. and S. Merz in München,* is still in good conditions, though altered in its assembly, and seems almost unused: this would be consistent with the fact that Secchi criticized it (Merz 1872a). The hypothesis that it could belong to the equipment of the Capitol Observatory—consequently having nothing to do with Secchi—cannot be excluded, but it is entirely to be explored (see below).

5.5 *Other Accessories*

In his book *Le Soleil*, Secchi also mentioned a Merz polariscopic eye-piece (Secchi 1875–1877, I, p. 36), but it is uncertain if this small instrument is still extant, as it has not yet been found in the depots of the Rome Observatory. Prisms, micrometers and other accessories were also certainly delivered by Merz to Secchi (see Wolfschmidt and Kost in this volume), but the merging of the historical collections of the Collegio Romano and the Capitol Observatories, as well as their move, in 1923, to the new Rome Observatory at Monte Mario and, later, to the new site at Monte Porzio, renders retrieval and identification of these items as quite difficult.[49]

6 Final Remarks

The role of the instrument-makers, in the history of science, is generally less emphasized than that of the scientists. However, their contribution has often been so important that it should not be ignored, being sometimes inextricable from the scientists' contributions. In his famous book *The history of the telescope*, when dealing with early astrophysicists, Henry C. King wrote:

[48]Secchi also possessed a direct-vision spectroscope by Tauber. After the establishment of the Società degli Spettroscopisti Italiani (Chinnici 2008), in fact, Secchi's main scientific partner was the astronomer Pietro Tacchini (1838–1905), who worked at the Palermo Observatory. Tacchini used a Merz telescope, almost identical to the one that was installed at the Collegio Romano Observatory, combined with a Tauber direct-vision spectroscope. Therefore, in order to have entirely comparable spectroscopic equipment, he suggested to Secchi the purchase of a Tauber spectroscope like the one he used in Palermo (Chinnici and Gasperini 2013, p. 418). The Tauber spectroscope, acquired by Secchi, is today preserved at Rome Observatory.

[49]A couple of direct-vision prisms—probably by Merz—have recently been found in the depot of the Rome Observatory, but their provenance from the German workshop has not yet been confirmed.

> To their own inventiveness was added the constructional skill of a number of opticians and instrument-makers. It is no exaggeration to say that the rapid rise of astrophysics was made possible by craftsmen like C. A. Steinheil and the Merz organization in Munich, John Browning, William Simms [...] These workers soon became acknowledged experts in the design and manufacture of high-grade prisms, spectroscopes, and auxiliary spectroscopic apparatus. (King 1955, p. 284)

This is clearly proven in the case of the Secchi-Merz relationship, in which the interaction between the two actors appears to have been crucial for the development of Secchi's spectroscopic studies. The analysis of the Merzs' letters thus reinforces "*the dependence of the astronomer on the optician*" (ibid., p. 288) and confirms the importance of their collaboration for the development of science.

References

Annali dell'Ufficio Centrale di Meteorologia Italiana 1879, I.

Calisi M. 1991, *Guida alla visita del Museo Astronomico e Copernicano di Roma*, Rome: Osservatorio Astronomico di Roma.

Calisi M. 2009. '6.S.08. Objective Prism' in I. Chinnici (ed) *Astrum 2009. Astronomy and Instruments,* Rome: Edizioni Musei Vaticani & Sillabe.

Chinnici I. 2008. "The Società degli Spettroscopisti Italiani: birth and evolution", Annals of Science 65, 3, 393–438.

Chinnici I., Gasperini A. 2013. *Alle origini dell'astrofisica italiana: il carteggio Secchi-Tacchini 1861–1877.* Florence: Fondazione Giorgio Ronchi.

Fraunhofer J. 1817, "Bestimmung des Brechungs- und Farbenzerstreuungs verschiedener Glasarten, in Bezug auf die Vervollkommnung achromatischer Fernröhre", *Denkschriften der München Akademie der Wissenschaften* 5, 193–226.

Hearnshaw, J. 1991, *The analysis of starlight,* Cambridge University Press.

Hearnshaw, J. 2009, *Astronomical spectrographs and their history,* Cambridge University Press.

Hentschel, K. 2002, *Mapping the Spectrum*, New York: Oxford University Press.

Hodgson, R. 1854, "Description of an Eye-piece for Observing the Sun", *Monthly Notices of the Royal Astronomical Society*, XV, 8 Dec 1854, No. 2, 45.

Huggins, W. 1864, "On the spectra of some of the Nebulae", *Philosophical Transactions of the Royal Society of London,* 154, 437–444.

King H. C., 1955, *The history of the telescope*, London: C. Griffin & Co. Ltd

Merz G. to A. Secchi, Munich, 21 July 1856, APUG, FS, 16.

Merz G. to A. Secchi, Munich, 29 January 1862 [a], APUG, FS, 16.

Merz G. and S. to Secchi, Munich, 7 June 1862 [b], APUG, FS, 16.

Merz G. and S. to Secchi, Munich, 19 January 1863 [a], APUG, FS, 16.

Merz S. to A. Secchi, Munich, 25 January 1863 [b], APUG, FS, 16.

Merz S. to A. Secchi, Munich, 11 January and 10 July 1864 [a], APUG, FS, 16.

Merz S. to A. Secchi, Munich, 29 April 1864 [b], APUG.

Merz S. to A. Secchi, Munich, 24 July, 24 September and 7 November 1864 [c], APUG, FS, 16.

Merz S. to A. Secchi, Munich, 29 December 1864 [d], APUG, FS, 16.

Merz S. to A. Secchi, Munich, 3 March 1865 [a], APUG, FS, 16.

Merz S. to Secchi, Munich, 15 November 1865 [b], APUG, FS, 16.

Merz S. to A. Secchi, Munich, 31 January 1866 [a], APUG, FS, 16.

Merz G. to A. Secchi, Munich, 18 September 1866 [b], APUG, FS, 16.

Merz G. to A. Secchi, Munich, 9 October 1866 [c], APUG, FS, 16.

Merz S. to A. Secchi, Munich, 17 June 1867 [a], APUG, FS, 16.

Merz S. to A. Secchi, Munich, 28 July 1867 [b], APUG, FS, 16.
Merz S. to A. Secchi, Munich, 7 September 1867 [c], APUG, FS, 16.
Merz S. to A. Secchi, Munich, 28 December 1867 [d], APUG, FS, 16.
Merz S. to A. Secchi, Munich 10 March 1868 [a], APUG, FS 16.
Merz S. to A. Secchi, Munich 16 March 1868 [b], APUG, FS 16.
Merz S. to A. Secchi, Munich, 27 March 1868 [c], APUG, FS, 16.
Merz S. to A. Secchi, Munich, 18 May 1868 [d], APUG, FS, 16.
Merz S. to A. Secchi, Munich, 2 June 1868 [e], APUG, FS, 16.
Merz S. to A. Secchi, Munich, 27 December 1868 [f], APUG, FS, 16.
Merz S. to A. Secchi, Vienna, 15 September 1869 [a], APUG, FS, 16.
Merz S. to A, Secchi, Munich, 6 March 1869 [b], APUG.
Merz S. to A, Secchi, Munich, 17 March 1869 [c], APUG.
Merz S. to A, Secchi, Munich, 11 May 1869 [d], APUG.
Merz S. to A, Secchi, Munich, 8 July 1869 [e], APUG.
Merz S. to A. Secchi, Munich 17 November 1869 [f], APUG, FS, 16.
Merz, S. 1870 [a]. "Objectiv-Spectralapparat", *Repertorium für Experimental-Physik* Bd. 6, 164–65.
Merz S. to A. Secchi, 6 June 1870 [b], APUG, FS, 16.
Merz S. to A. Secchi, Munich, 17 October 1871, APUG, FS, 16.
Merz S. to Secchi, Munich 14 and 29 February 1872 [a], APUG, FS, 16.
Merz, S. 1872 [b]. *Verzeichniss der Instrumente welche in dem Optischen Institute von den G. und S. Merz vormals Untzschneider und Fraunhofer in München,* Druck von J. G. Weiss, Universitatsbuchdrucker.
Merz S. to A. Secchi, Munich, 27 May 1873 [a], APUG, FS, 16.
Merz S. to A. Secchi, Munich, 19 July 1873 [b], APUG, FS, 16.
Merz S. to A. Secchi, Munich, 16 March 1874 [a], APUG, FS, 16.
Merz S. to A. Secchi, s. l., 2 May 1874 [b], APUG, FS, 16.
Merz S. to A. Secchi, s. l., 6 February 1876 [a], APUG, FS, 16.
Merz S. to A. Secchi, s. l., 9 June 1876 [b], APUG, FS, 16.
Morris P. J. T. (ed.) 2010, *Science for the Nation. Perspectives on the History of the Science Museum,* London: Science Museum.
Secchi, A. 1855. "Sur un nouveau système de micromètres pour les lunettes astronomiques", *Comptes-Rendus de l'Académie des Science,* XLI, pp. 906–907.
Secchi A. 1856. *Descrizione del nuovo Osservatorio del Collegio Romano e Memoria sui lavori eseguiti dal 1852 a tutto aprile 1856*, Roma: Tipografia delle Belle Arti.
Secchi, A. 1867. « Sugli spettri prismatici delle stelle fisse » , *Memorie della Società Italiana delle Scienze*, Serie III, Tomo I, Parte I, pp. 67–152.
Secchi A. 1869. « Sugli spettri prismatici delle stelle fisse, Memoria II » , *Memorie della Società Italiana delle Scienze*, Serie III, Tomo II, pp. 1–60.
Secchi A. 1870, *Le Soleil*, Paris: Gauthier-Villars.
Secchi, A. 1875–77, *Le Soleil*, 2nd edn, 2 vols, Paris: Gauthier-Villars.
Secchi A. 1895, *Les Etoiles*, 3rd edn, 2 vols, Paris: F. Alcan.
Struve F. G. W. 1845. *Description de l'Observatoire astronomique central de Poulkova,* St. Petersburg: Imprimerie de l'Académie Impériale des Sciences.
Udìas A. 2003. *Searching the Heavens and the Earth: the History of Jesuit Obervatories*, Dordrecht: Kluwer Academic Publshers.
Warner D. J., 'Direct vision spectroscopes', *Rittenhouse* (1993) 7, pp. 40–48.

The Padua Observatory and the Merz Workshop Under the Austro-Hungarian Empire

Valeria Zanini

The Padua Astronomical Observatory operated under the Government of the Austro-Hungarian Empire from 1813 to 1866 and consequently, in those years, had a special connection to such German optical firms as Utzschneider and Fraunhofer and with their successor, Merz.

The Museum *La Specola*, belonging to the Padua Observatory, keeps several instruments whose optical parts were molded under Merz's supervision: an achromatic telescope (85 mm aperture) and a 'parallactic machine' (i.e., an equatorial telescope), both made in 1822 and signed 'Utzschneider u. Fraunhofer'; a 'comet seeker' and an altazimuth telescope commissioned by Giovanni Santini in 1842, on the occasion of the 4th Congress of Italian Scientists held in Padua; an equatorial telescope (116 mm aperture) dating back to 1858, and the large equatorial (187 mm aperture), constructed in 1862 and moved to Padua in 1881, that once belonged to Baron Ercole Dembowski. The history and description of these instruments is given below.

1 The First German Instruments at the Padua Observatory

On May 12th, 1797, Ludovico Manin (1725–1802), the last Doge of *La Serenissima* (the Venetian Republic), surrendered unconditionally to the French Napoleonic troops and abdicated, while the Major Council declared the end of the millenary Republic, under which the Padua Astronomical Observatory had been founded just thirty years before. This event opened up a long period of political and administrative instability for the city of Padua, as well as for all territories belonging

V. Zanini (✉)
INAF-Osservatorio Astronomico di Padova, Padova, Italy
e-mail: valeria.zanini@oapd.inaf.it

I. Chinnici (ed.), *Merz Telescopes*, Historical & Cultural Astronomy,
DOI 10.1007/978-3-319-41486-7_4

to *La Serenissima*: from 1797 to 1806, there was a frequent swap between French and Austrian troops, which alternately maintained control of the city. Finally, in 1806, Padua found itself under a stable government, the Italic Kingdom, but it was short-lived; in 1813, in fact, the city fell under the domain of the Austro-Hungarian Empire, which ruled the ex-Venetian lands until 1866, when the whole region was annexed to the newly formed Kingdom of Italy, after the Italian *Risorgimento*.

Thus began a long period of political stability for Padua, as well as for the Astronomical Observatory; this coincided with the long directorship of Giovanni Santini[1] (1787–1877). Thanks to the administrative support that the new Government ensured, Santini could equip the Observatory with new modern instrumentation. Until then, the Observatory had turned to local or British instrument-makers for the supply of its scientific equipment, but, once under the aegis of the Austro-Hungarian Empire, business relationships with German and Austrian companies were quite logically facilitated. In particular, Santini turned immediately to the 'Reichenbach and Utzschneider' firm to purchase (Zanini 2007a) a transit instrument, given the Observatory's lack of one, and later, in 1815, a repeating circle[2] (Fig. 1).

The instrument-maker Georg Friedrich von Reichenbach (1772–1826) and the businessman Joseph von Utzschneider (1763–1840) had been partners with the famous optician Joseph Fraunhofer (1787–1826) since 1809. In 1808, the young Georg Merz (1793–1867) started working in the Utzschneider workshop and soon displayed great skill, so that he became, later, "*Er wurde Vorarbeiter der Glasschleiferei und Fraunhofer's Amanuensis*" ("*foreman of glass grinding and Fraunhofer's amanuensis*") (Merz 1868, p. 363). When these early Paduan instruments made in Germany were constructed, Merz was already Fraunhofer's assistant and he most likely participated directly in the manufacture of the optics.

A few years later, Santini turned back to Utzschneider for a modern parallactic machine. In 1816, he had applied to the competent authority of the University for the funds needed to purchase this kind of instrument, not only because the Observatory lacked such a telescope, but also because, in the previous year, he had been invited by the German astronomer Bernard von Lindenau (1779–1854), from the Gotha Observatory, to take part in an international project for "*una minuta e precisa revisione del cielo*" ("*a minute and thorough survey of the sky*") (HAOPd 1816; see also HAOPd 1815), planned by Heinrich Wilhelm Olbers (1758–1840),

---

[1]Santini, a young brilliant astronomer, former pupil of the Milan astronomers Barnaba Oriani (1752–1832) and Angelo De Cesaris (1749–1832), was both a practical and theoretical astronomer. His most important and demanding astronomical work was a catalogue of stars between declination +10° and −10°, which he completed in 1857, after twenty years of observations made with the meridian circle that he purchased in 1837. He acquired his greatest repute by his calculations of comets' orbital periods (Pigatto 1996).

[2]This instrument is used by 'repeating' several observations of the angular distance of two objects and of their zenithal distance. For each observation, it is possible to obtain increasing integer multiples of the unknown angle, thus reducing reading errors and obtaining more precise measurements (Zanini 2007b, p. 158).

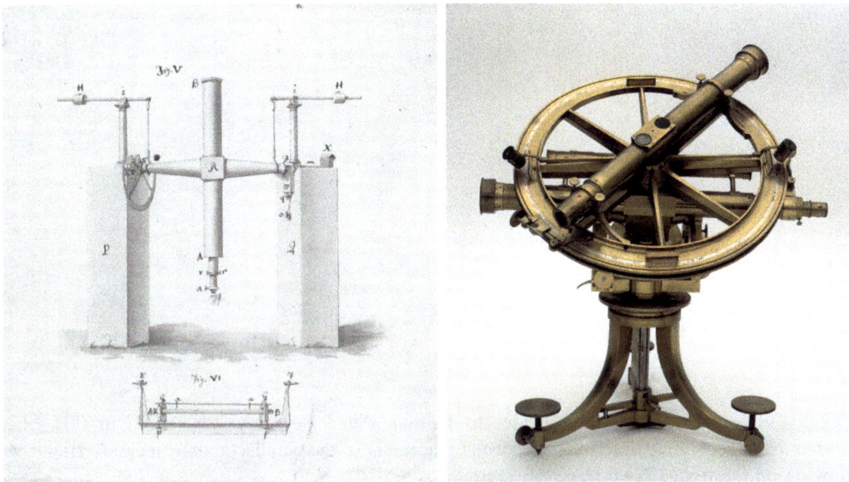

Fig. 1 *On the left*: the old transit instrument made by Reichenbach and Utzschneider in 1811, no longer preserved among the ancient instruments of the Observatory. *On the right*: the repeating circle made in 1815 by Reichenbach (INAF-HAOPd)

Carl Friedrich Gauss (1777–1855) and Lindenau himself. Such a work, however, could not be carried out without a precision instrument. Santini was very satisfied with both the transit instrument and the repeating circle made by Reichenbach. In particular, he praised "*l'esimia riuscita*" ("*the remarkable performance*") (HAOPd 1816) of the latter instrument and, consequently, he wished to acquire another precision instrument from this firm.

Unfortunately, Santini's request was rejected by the University and he was forced to give up the collaboration proposed by Lindenau. A few years later, when he officially became director of the Observatory, he again applied for the purchase of this instrument, which was absolutely necessary for his studies on comets, the celestial objects of greatest interest to him. This time, the request was granted; however, the instrument was not delivered to Padua observatory until 1822; it was accompanied by detailed instructions from Utzschneider for its correct placement (Fig. 2). Certainly, Georg Merz supervised the construction of the optics for this instrument, as he had become foreman of the Benediktbeuern factory in the previous years.

The circles of right ascension and declination of this parallactic machine have a diameter of 24 Parisian inches (64 cm) and are silver graded. Both circles are equipped with a pair of vernier scales that provide an accuracy of a second and four arc-seconds, respectively. The telescope, 29-lines (about 6.5 cm) in aperture and 30 Parisian inches (80 cm) in focal length, is pivoted in the centre of the declination circle.

This instrument replaced a poor quality and antiquated equatorial telescope made by Gian Battista Rodella (1749–1834), the first craftsman of the Observatory. It was

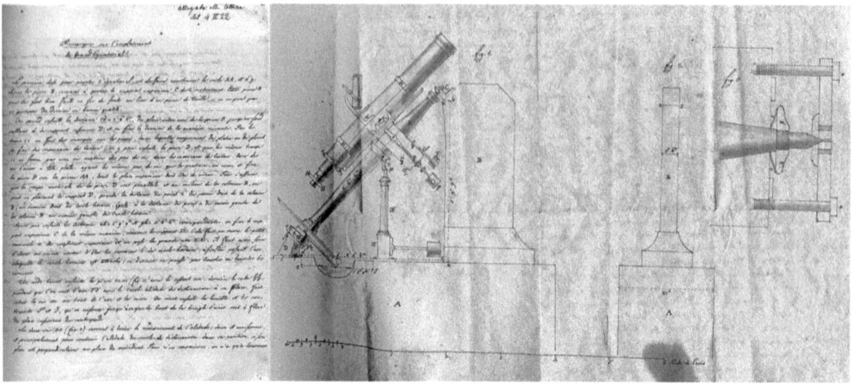

Fig. 2 Original letter by Utzschneider to Santini, dated February 4th, 1822; in this letter, Utzschneider accurately describes the correct placement of the parallactic machine, referring to the drawing on the right (INAF-HAOPd)

installed on the top of the tower, in the little dome specifically designed for this type of telescope. The Utzschneider machine was "*appoggiata ad una colonna verticale robusta di ferro fuso eretta sopra solidissima base di pietra istriana* ("*put in leaning on a strong cast-iron column, erected on a solid base of Istrian stone*") (Santini 1846, p. 133–footnote 1). It was used until 1858, when a new parallactic machine, the Starke-Merz Equatorial (see below), was acquired. Today, the old parallactic machine by Utzschneider, Reichenbach and Fraunhofer requires an adequate cleaning. A modern iron support has been recently constructed and the telescope is now waiting for a suitable placement within the museum exhibition of the Observatory (Fig. 3).

In 1822, Santini also purchased a second telescope, later called the 'Fraunhofer' (Fig. 4): it is an altazimuth instrument with an aperture of 37.5 lines (85 mm) and a focal length of 48 Parisian inches (130 cm). The supporting table, triangular in shape, was made expressly by the Observatory mechanical workshop in 1823. Unfortunately, the objective lens got broken in the 1970s, due to an incorrect use of the instrument.

The telescope was brought by Giuseppe Lorenzoni (1843–1914)[3] to Sicily in 1870, for observing the solar eclipse visible from the south-eastern part of the island on December 22nd. On that occasion, it was used as the seeker for the Starke-Merz

[3]Lorenzoni was the fourth director of the Padua Astronomical Observatory. His scientific activity ranged from classical astronomy to spectroscopy and geodesy. He introduced solar physics research to Padua Observatory, and, in 1871, was one of the promoters of the "Società degli Spettroscopisti Italiani" (Italian Spectroscopist Society), the first "astrophysical" society, which created a network among astronomers working in solar physics in Italy and abroad. He was, above all, an excellent teacher, who had inspired almost an entire generation of astronomers by the end of the 19th century; under his leadership, the Paduan school of astronomy became a symbol of excellence for astronomical education in Italy (Zanini 2015).

Fig. 3 The Utzschneider parallactic machine, in its current state (from: Zanini 2007b, p. 150)

Equatorial, purchased in 1858. In 1998, the instrument was rediscovered by chance at the Asiago branch of the Padua Observatory; while the objective-lens was irreparably broken, its mechanical parts were later restored and the telescope is now on display in the museum.

2 Telescopes with Merz Optics Purchased Through the Polytechnic Institute of Vienna

Another important purchase for the Observatory was the great meridian circle (Fig. 5) used by Santini to carry out his famous *Paduan Catalogues* (Zanini 2007a and Zanini 2007b, p. 156). The entire mechanical structure of this instrument was constructed by Christoph Starke (1794–1865) at the Polytechnic Institute of

Fig. 4 The achromatic telescope made by Utzschneider and Fraunhofer (from: Zanini 2007b, p. 149)

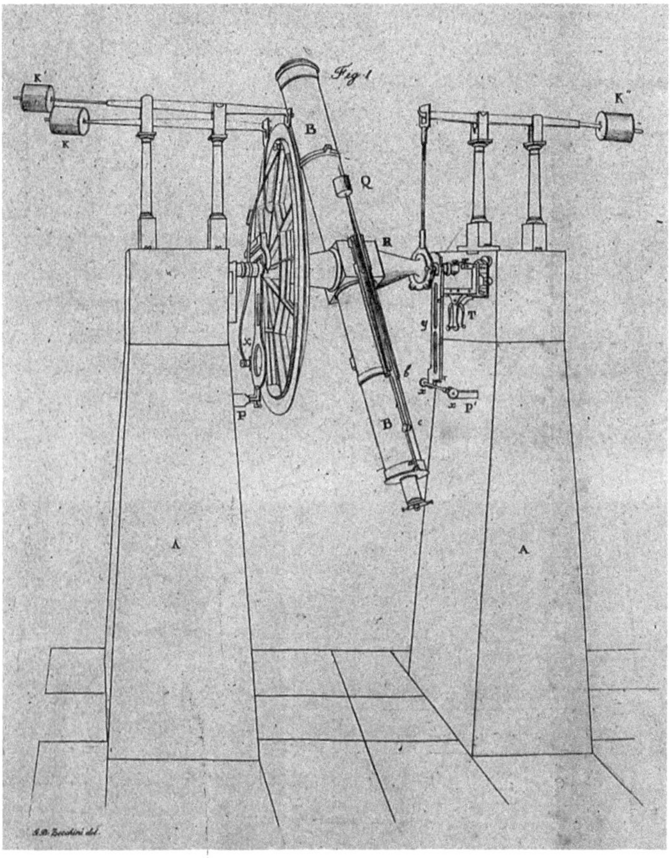

Fig. 5 Drawing of the meridian circle, made by Christoph Starke for the Padua Astronomical Observatory in 1836 (INAF-HAOPd)

Vienna,[4] but its optical elements were provided by the Fraunhofer firm, namely by Georg Merz, who had inherited the signature. Unfortunately, in 1911, the original Merz objective-glass was replaced by a Zeiss lens and the old one, along with several other instruments, was traded to the Italian Government and used during the First World War.

Up through the time of the meridian circle acquisition in the 19th century, all of the Padua Observatory's purchases of new optical equipment from the Merz firm were accomplished via Christoph Starke, who constructed the mechanical parts.

[4]The Imperial and Royal Polytechnic Institute of Vienna, now the Vienna University of Technology, was founded in 1815. It was, at once, an 'academy' for all branches of arts, trades and commerce; a 'conservatory', in which instruments, machines and productions of all the arts were assembled; and a 'society for the encouragement of national industry', which proposed and distributed prizes and awards.

This is the case for the instruments purchased by Santini in 1842, an altazimuth telescope and a comet-seeker, both by Starke, which were acquired on the occasion of the Fourth Meeting of Italian Scientists, held in Padua: Santini took advantage of that circumstance to renew the equipment of the Observatory and asked the Austrian Government to purchase them, in order to give a good impression about the Observatory to the eminent visitors.

The Starke altazimuth telescope (Fig. 6, *on the left*) is fixed to a wooden tripod laterally, and its equilibrium is supplied by a counterweight. On the azimuth circle, which is fixed at the top of the tripod, is engraved the signature *K.K. polÿtechn. Institut in Wien. Chr. Starke*. The 117-mm aperture objective-glass, made by Utzschneider and Fraunhofer, has a 195-cm focal length; the telescope is equipped with an achromatic seeker, 3-cm aperture and 30-cm focal length. The tripod is provided with wheels, so that the whole instrument may be freely moved. This refractor was shipped to India in 1874 to observe the rare event of the transit of

Fig. 6 Instruments made at the Polytechnic Institute in Vienna, with Merz optics. *On the left*: The Starke altazimuth; *On the right*: the Merz comet seeker

Venus across the solar disc (Pigatto and Zanini 2001; Chinnici 2003). For many years, the telescope was left abandoned in the tower of the Observatory, under very poor conditions; it was only restored in 1995.

The second telescope acquired in 1842, the Starke comet-seeker[5] (Fig. 6, *on the right*), has an equatorial mounting, a wooden tube and brass housings for the objective-glass and the eye-piece. The objective-lens, 77-mm aperture and 650-mm focal length covers a field of six degrees. This time, it is the mounting that is engraved *K.K. polÿtechn. Institut in Wien. Chr. Starke*, while the eye-piece bears *Utzschneider u. Fraunhofer in Munchen*, the original signature, initially unchanged by Merz. Santini was well-known because of his very precise computations of cometary orbits, but he aspired to discover new ones: unfortunately for him, he never succeeded, despite having purchased this instrument for that very purpose. This telescope, restored in 1995, is now on display at *La Specola* Museum.

3 Santini's Journey to Germany

In March 1843, a tragic event struck Santini: his beloved wife, Teresa Pastrovich (1778–1843), whom he married in 1810, died after a long and painful illness, leaving her husband in a state of deep despair. To distract his mind from this terribly sad event, in the ensuing autumn, Santini engaged in a trip he had long planned but never attempted: "*un viaggio per la colta Germania ad oggetto di conoscere da vicino i celebri Osservatorj di Vienna, Berlino, Amburgo, Altona e Monaco*" ("*a journey through the cultured Germany in order to get to know the famous Observatories of Vienna, Berlin, Hamburg, Altona and Munich*") (Santini 1846, p. 142).

Travelling through today's Austria, Czech Republic and Germany, between September and October of 1843, he visited the astronomical Observatories of several cities: Vienna, Prague, Leipzig, Halle, Berlin, Altona, Hamburg, Regensburg and Bogenhausen. He also visited workshops and factories where astronomical instruments were produced: Starke's workshop and the optical institute of Simon Plössl (1794–1868) in Vienna; the firms of Christian Tiede (1794–1877), Carl Pistor (1778–1847) and Johann Oertling (1803–1866) in Berlin; the workshop of the Repsold brothers, Georg (1804–1885) and Adolph (1806–1871), in Hamburg, and, finally, in Munich, those of Traugott Leberecht Ertel (1778–1858)—successor to Reichenbach's firm—and Georg Merz with his son Sigmund (1824–1908).

Regarding the Merz workshop, from which all the optical parts of the telescopes he had acquired had come, on October 11th, 1843, Santini noted in his diary:

[5]Comet-seekers are telescopes that have typical luminosity and a wide field and are easy to put into operation; for these reasons, they are suitable for observing faint objects, such as comets, in any part of the sky.

> *I found father and son both courteous. I saw three giant objectives, one of them, with a 14-inch aperture, was destined for Pulkovo. The workshop was not active, because it was night. They showed me the machinery for grinding lenses [...]. They told me that Fraunhofer had the same equipment, which is somewhat difficult for me to believe. I asked if they work the eye-piece lenses separately or jointly: they told me separately, as they are unable to work them jointly. This answer is the opposite to that which I received in Vienna from Plössl, who works these lenses together, six or eight at a time, and who told me that, if separated, they are less regular and less perfect.*[6]

Santini returned to Italy reinvigorated and full of new ideas. In order to support observations, he introduced a lot of innovative devices discovered during his journey, such as the micrometric microscopes, with which he replaced all the vernier scales of the meridian circle and the other telescopes of the Observatory.

4 A New Telescope for the Observatory

A few years later, Santini purchased from Starke a new equatorial telescope, equipped with Merz optics. Santini required that it had an objective aperture comparable in size to that of the meridian circle and, above all, he wanted an instrument with the «gear» to drive the machine "*con moto uniforme, e senza scosse nel senso del moto diurno degli astri*" ("*with uniform motion, and without vibrations, in the direction of the diurnal motion of the stars*") (HAOPd 1852).

This was the first telescope with a clockwork mechanism in the equipment of the Padua Observatory. However, it took six years to be manufactured before it could be used by the Paduan astronomers. Its construction, in fact, was delayed for many years, because Starke, as foreman of the Polytechnic Institute's mechanical workshop, "*contro sua voglia per espresso volere del Ministero dovette prender parte all'Esposizione di Parigi colla costruzione di molti strumenti*" ("*against his desire, by direct order of the Ministry, had to take part in the Exposition Universelle [International Exhibition] held in Paris on the design of many instruments*") (HAOPd 1855). The exhibition took place in spring 1855 and, after its conclusion, Starke was able to start construction of the Equatorial telescope for the Padua Observatory, to which the instrument was delivered in 1858.

The declination and right ascension circles of this telescope have a diameter of 27.5 cm and are silver graded. Both circles are equipped with a pair of vernier scales

---

[6]*Trovai il padre ed il figlio ambidue cortesi. Vidi tre giganteschi obbiettivi, uno dei quali di 14 pollici di apertura destinato per Pulkova. L'officina non era più in attività perché era di notte. Mi mostrò la macchina a cui arrotano le lenti [...]. Mi dissero che Fraunhofer aveva gli stessi apparati e quasi stenterei a crederlo. Domandai se le lenti oculari le lavorano separate o unite: mi dissero separate, non riuscendo di lavorarle unite. Risposta contraria di quella che ottenni in Vienna da Plössl, il quale le lavora unite a sei od otto per volta, e dichiarommi che, isolate, riescono meno regolari e meno perfette.* Copy of the original in Lorenzoni (1887), pp. 167–168. Unfortunately, Santini's original diary appears to be missing from the Historical Archive of the Padua Astronomical Observatory.

giving it accuracy on the order of half a second and four arc-seconds, respectively. The tracking mechanism (now lost), in association with the equatorial mounting, provided the rotational diurnal motion thanks to the action of a variable number of weights. By increasing or decreasing them, it was also possible to render the angular speed of the telescope equal to that of the Sun, the planets and the Moon. The entire mechanical structure is supported by a molten iron column with a three-leg base. The size of the Merz objective, with a 116-mm aperture and 1640-mm focal length, is exactly as requested by Santini. It was installed in 1861 in a cylindrical room with a rotating dome, built on top of the observatory tower (Fig. 7).

The telescope was used primarily for observations of comets and asteroids (Fig. 8), but from 1870 to the 1880s, it was also used by Lorenzoni for

Fig. 7 The Merz-Starke equatorial in its original dome at the top of the observatory tower, in a photograph from the late 1940s (INAF-HAOPd)

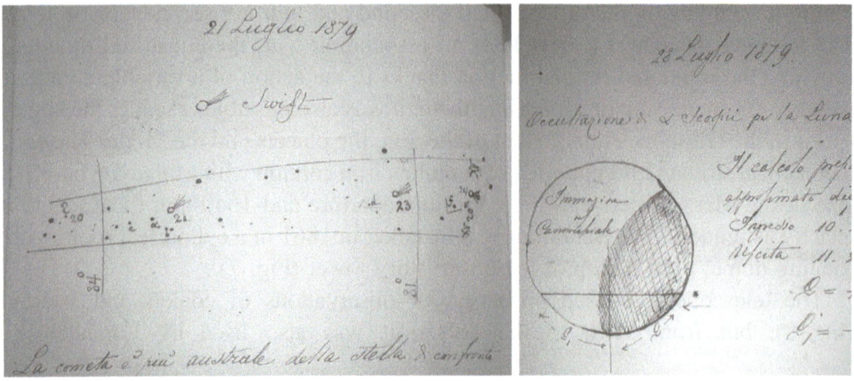

Fig. 8 Astronomical observations made by Antonio Abetti (1846–1928) in July 1879 with the Merz-Starke equatorial. *On the left*: observation of the comet (C/1879 M1) discovered by Lewis Swift (1820–1913); *on the right*: Lunar Occultation of α Scorpii

Fig. 9 The Merz-Starke equatorial today, restored and on display at the Museum *La Specola* (from: Zanini 2007b, p. 152)

astrophysical studies: in fact, in combination with a Hofmann direct vision spectroscope, it became the main research instrument for solar physics observations at the Padua Observatory (Zanini 2010). This telescope was also part of the scientific equipment used by the previously-mentioned Italian astronomical expeditions to observe the total solar eclipse of 1870 and the transit of Venus of 1874.

Up until the end of the 20th century, the telescope was left unused in its dome at the top of the tower; it was restored in the 2000s and is currently on display at *La Specola* Museum (Fig. 9).

5 An Important Addition: Dembowski's Merz Telescope

The Observatory acquired its last telescope with Merz optics in 1881. Padua had been freed from the Habsburg Empire's domination and annexed to the young Kingdom of Italy just fifteen years before; Giuseppe Lorenzoni had succeeded Giovanni Santini (who died in 1877) as director of the Observatory. Times had changed, but this new purchase, however, was not the purchasing of a new and innovative instrument, rather it was the acquisition of a 'second-hand' used telescope. The new Italian State was facing severe economic and social problems and allocated scarce or null financial resources for scientific research; consequently, the Italian astronomers could not purchase astronomical instruments comparable to those of the best European and overseas Observatories.

This 'new' telescope belonged to Baron Ercole Dembowski (1812–1881),[7] an amateur astronomer who had established a private observatory "*da fare invidia a più d'uno Osservatorio Governativo*" ("*rivalling more than one Government Observatory*") (HAOPd 1881). For many years, he devoted himself to the study of double stars, obtaining a good international reputation (Dembowski 1864). After the Baron's death, his heirs sold all his instruments to the Italian Government, which destined this telescope, the most important of Dembowski's instruments, specifically to the Padua Observatory.

The objective-glass of this telescope (Fig. 10), made by Merz in 1862, has an 187-mm aperture and a 3200-mm focal length. It was originally equipped with several eyepieces, which provided magnifications from 100 to 720 times, and with a bad clockwork mechanism for tracking stars, now lost (Dembowski 1864,

[7]Baron Dembowski, son of one of Napoleon's Polish Generals and his Italian wife, spent several years in a military career in the Navy, before retiring to private life. He was passionate about astronomical studies and, originating from a rich family, was able to establish an excellent private observatory at S. Giorgio a Cremano, near Naples; later, he decided to move the observatory to Gallarate, near Milan.

Fig. 10 The 'Dembowski' telescope in its original dome in the first half of the 20th century (INAF-HAOPd)

pp. 129–130). The right ascension circle is divided into minutes, whose reading is made by means of a vernier scale, whereas the declination circle, provided with another vernier scale, is divided into intervals of ten arcminutes. This telescope was the main observing instrument for the Paduan astronomers until the establishment of the Asiago Astrophysical Observatory in 1942, and it was primarily used for observations of planets, comets and asteroids (Fig. 11).

By the late 1970s, the telescope was no longer being used and was disassembled; only the Merz objective-lens was preserved with some care, to be used with another mounting (see Massone, in this volume). It became possible to carry out a restoration of the whole telescope during the International Year of Astronomy (2009); the restored telescope was publicly displayed, for the first time, at the *Astrum* 2009 exhibition, held in Rome at the Vatican Museums (Astrum 2009; Pigatto 2014). Nowadays, this is the main instrument of the Museum *La Specola* collection (Fig. 12).

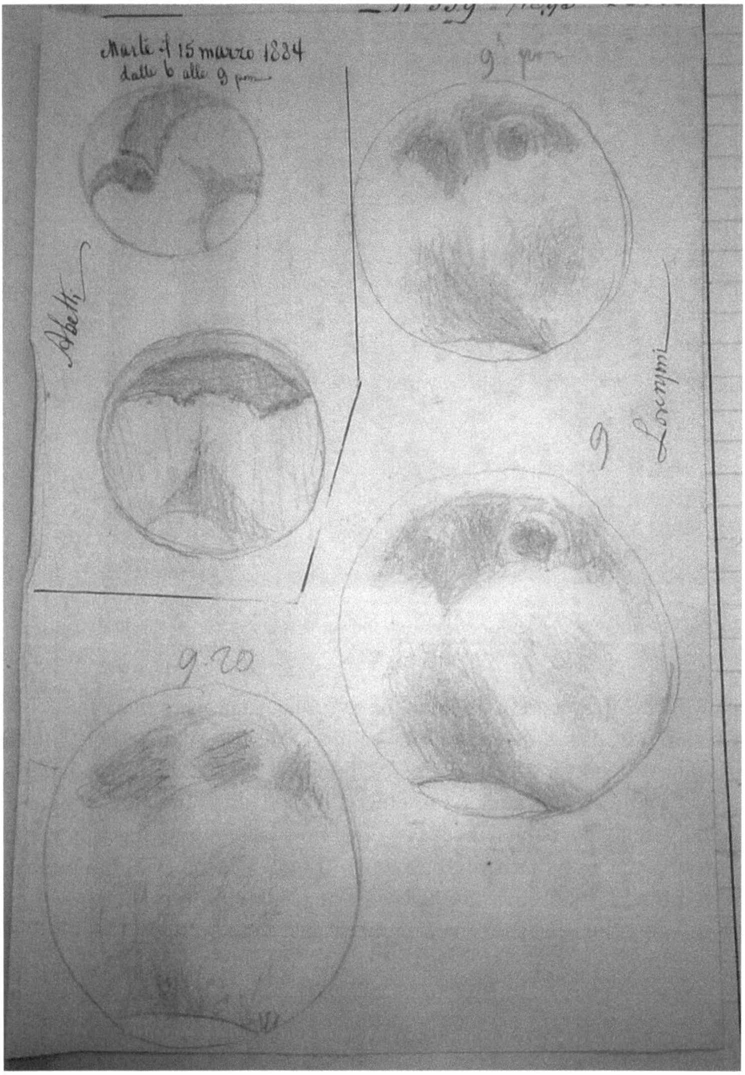

Fig. 11 Mars observations made by Antonio Abetti and Giuseppe Lorenzoni in 1884 with the 'Dembowski' telescope (INAF-HAOPd)

Fig. 12 The 'Dembowski' telescope restored and on display at the observatory museum

It is worthwhile mentioning that the original rotating dome, built by Baron Dembowski, was also acquired by the Padua Observatory, together with the telescope; a plot of land on private property, between the Bacchiglione River and the 'Naviglio' Canal, was later rented to erect a walled structure supporting the dome. Occasionally renovated over the years, but now abandoned for a long time, the dome has become rusty and is falling down (Fig. 13). However, having been built on private property, its repair and subsequent restoration for educational purposes and the safeguarding of history is a problem that may be very difficult to solve.

Fig. 13 The original dome of the 'Dembowski' telescope, in its present condition

References

Astrum 2009: *Astrum 2009. Astronomy and instruments: Italian heritage four hundred years after Galileo*. I. Chinnici Editor, Musei Vaticani - Livorno, Sillabe, 2009.

Chinnici 2003: *Transito di Venere 1874: una spedizione italiana in Bengala*, «Giornale di Astronomia», 2003, 4, pp. 45–53.

Dembowski 1864: *Beobachtung von Doppelsternen, angestellt auf der Sternwarte des Herren Barons Dembowski*, «Astronomische Nachrichten» nr 1473–1475 (1864), pp. 129–170.

HAOPd (Historical Archive of the Astronomical Observatory of Padua) 1815: *Fondo Santini - Correspondence*, b. III, Letter of Lindenau to Santini, December 1st, 1815.

HAOPd 1816: *Atti e dati relativi a Strumenti*, b. XV, f.1, Draft letter of G. Santini to Pietro di Goess, Padua, May 6th, 1816.

HAOPd 1852: *Atti e dati relativi a Strumenti*, b. XV, f. 1, Draft letter of G. Santini to University Rector, w.d. (but 1852).

HAOPd 1855: *Atti e dati relativi a Strumenti*, b. XV, f. 1, Letter of F. Rossetti to G. Santini, May 24 h, 1855.

HAOPd 1881: *Atti e dati relativi a Strumenti*, b. XV, f. 6, Draft report for the purchase of 'Dembowski' equatorial, February 2nd, 1881.

Lorenzoni 1887: *In occasione del primo centenario dalla nascita dell'astronomo Santini*, «Atti e Memorie della R. Accademia di scienze lettere ed arti in Padova», Nuova Serie, Vol. III (1887), pp. 133–183.

Merz 1868: *Kurzer Lebensahriss von Georg Merz*, «Astronomische Nachrichten», nr 1679 (1868), pp. 361–364.

Pigatto 1996: *Giovanni Santini*, in *Professori di materie scientifiche all'Università di Padova nell'Ottocento*, Casellato-Pigatto Editors, Trieste 1996, pp. 35–40.

Pigatto 2014: *Il 'Dembowski'. Storia di un cannocchiale*, «Giornale di Astronomia», 2014, 2, pp. 28–34.

Pigatto-Zanini 2001: *Spectroscopic observations of the 1874 transit of Venus: the Italian party at Muddapur, east India*, «Journal of Astronomical History and Heritage», 2001, 4, pp. 43–58.

Santini 1846: G. Santini, *Osservazioni intorno alle comete apparse nell'anno 1843*, «Memorie di Matematica e di Fisica della Società Italiana», XXIII, pars matematica, Modena 1846, p. 132–170.

Zanini 2007a: *Gli strumenti degli astronomi alla Specola di Padova*, in «Le scienze astronomiche nel Veneto dell'Ottocento. Atti dell'ottavo seminario delle Scienze e delle Tecniche», M.C. Ghetti Editor, Venezia, Istituto Veneto di SS.LL.AA., 2007, pp. 141–167.

Zanini 2007b: *Gli strumenti della Specola. Catalogo*, in L. Pigatto, *La Specola di Padova. Da torre medievale a museo*, Padova, Signum Editore, 2007, pp. 147–174.

Zanini 2010: *Lo spettroscopio a visione diretta e le ricerche astrofisiche sul finire dell'800 all'Osservatorio Astronomico di Padova*, «Giornale di Astronomia», 2010, 1, pp. 13–20.

Zanini 2015: *Giuseppe Lorenzoni: l'uomo, l'astronomo e il maestro*, «Atti e Memorie dell'Accademia Galileiana di Scienze Lettere ed Arti già dei Ricovrati e Patavina», Pars II: 'Memorie della Classe di Scienze Matematiche, fisiche e naturali', CXXVII, 2015, pp. 73–103.

The Merz Refractors at the Brera Astronomical Observatory

Mario Carpino

After its establishment at the end of the eighteenth century, the Brera Observatory gained international renown thanks to the reputation of its directors, such as Louis Lagrange (1711–1783) and Ruggiero Boscovich (1711–1787). During the second third of the nineteenth century, however, it met with a period of progressive decadence, which was triggered by the deterioration of the political relationships between Lombardy and the Austrian government, after the end of the Napoleonic era and the Congress of Vienna. As to the three astronomers working in Brera in 1815, Barnaba Oriani (1752–1832) retired in 1817 and Angelo de Cesaris (1749–1832) died in 1832, and since neither one of them was ever replaced, in 1860, only the elderly Francesco Carlini (1783–1862) remained. There was, consequently, obsolescence in the instrumental equipment, with the notable exceptions of the acquisitions of a Starke meridian circle in 1832 and an Amici 32 cm reflector in 1839. The new director, Giovanni Virginio Schiaparelli (1835–1910), described the situation he found in Brera in 1860 by stating that:

> … in the decade from 1850–1860, almost nothing remained [*of the Brera Observatory*] but the memory of its past splendor, and the name of an eminent old man [*Carlini*]. From 1830 on, the Brera astronomical studio no longer attracted the sort of men as distinguished as those who had rendered it famous up until then.[1]

An event revealing the climate of the epoch is the bequest made by Barnaba Oriani to the Observatory in 1832, as Schiaparelli commented:

> Oriani, in order to prevent everything [*at the Observatory*] from falling into ruin, made a testamentary disposition of 200 thousand Austrian liras in favour of the Observatory so that an annual pay of 4500 Austrian liras could be given to the second astronomer and 900 liras to a third scholar. There is every reason … to believe that, without this action by Oriani, the

[1]… nel decennio 1850–1860 altro quasi non ne rimaneva che la ricordanza degli antichi fasti, e il nome di un vecchio illustre. E dal 1830 in poi non uscirono più dallo studio astronomico di Brera tanti uomini distinti, e simili a quelli che lo avevan reso celebre fino a quell'epoca.

M. Carpino (✉)
INAF-Osservatorio Astronomico di Brera, Milano, Italy
e-mail: mario.carpino@brera.inaf.it

I. Chinnici (ed.), *Merz Telescopes*, Historical & Cultural Astronomy,
DOI 10.1007/978-3-319-41486-7_5

Observatory would no longer exist or it would have been abandoned, as happened with some others in Italy.[2]

The situation changed in 1860, again as a consequence of political events. After the defeat of the Austrians at the battle of Solferino (on June 24th, 1859) and the armistice of Villafranca, Lombardy was annexed to the Kingdom of Sardinia; soon, the new government, mainly under suggestion of Quintino Sella (1827–1884) and Luigi Menabrea (1809–1896), sent a new astronomer to Brera, who would play a determinant role in the history of the Observatory for the next four decades: Giovanni Virginio Schiaparelli. The Observatory also owes the acquisition of its three Merz refractors to him.

1 Schiaparelli and the Acquisition of Three Merz Telescopes

Schiaparelli was born on March 14, 1835 in Savigliano, near Cuneo (Piedmont). He was a member of a family which had acquired significant merit in the political and scientific fields: Cesare Felice (1821–1916), his uncle from his mother's side, was a general in the Savoyard army and participated in the Wars of Independence, earning several decorations; Eugenio, Giovanni's brother, was also a military man in the Savoyard army and died as the result of wounds received in the battle of Solferino in 1859; Celestino (1841–1919), another brother, was a scholar in Arabic literature. Giovanni Schiaparelli studied hydraulic engineering and architecture at Turin University, also following the lectures of the astronomer Giovanni Plana (1781–1864), and graduated in 1854. After teaching mathematics for some years in a middle school, he decided to devote himself to astronomy and joined a study group to tour the observatories of Berlin, where he collaborated with Johann Franz Encke (1791–1865), and Pulkovo, where he worked with Otto W. Struve (1819–1905). He was still in Pulkovo at the end of 1859 when he was appointed *secondo astronomo* at Brera Observatory, where he arrived in May 1860. Two years later, after the death of Francesco Carlini (on August 29th, 1862) he became *primo astronomo* (namely director): he was 27 years old, the youngest director in the history of the Brera Observatory. He then had the opportunity to fulfill a project for which he had quite likely been yearning for some time: to acquire a new, state-of-the-art instrument for the Observatory that would allow him to continue the astrometric studies which he had carried out at Berlin and Pulkovo, especially the measurement of binary stellar systems.

---

[2]Oriani, perché non andasse a fascio ogni cosa, dispose nel suo testamento di lire austriache 200 mila a favore della Specola per dare un soldo annuo di austriache 4500 ad un secondo astronomo e lire 900 ad un terzo allievo. E veramente … si ha ragione di credere, che, senza quest'atto di Oriani, l'Osservatorio più non esisterebbe o sarebbe abbandonato, come è avvenuto di alcun altro in Italia.

1.1 The First Merz Refractor at Brera

For his astronomical studies, Schiaparelli chose a Merz refractor (21.8 cm aperture, with a focal length of 315 cm, f/14.5), probably because he had used a similar instrument at Pulkovo. His request for the necessary funds from the Italian Parliament (the Kingdom of Italy had been proclaimed on March 17th, 1861) was backed up by his personal scientific reputation (mainly based upon his discovery of a new asteroid, Esperia, on April 29th, 1861) and was approved very quickly, thanks also to the support of Quintino Sella, who was Minister of Finances in those years. The instrument was ordered from Merz in 1862 and arrived in Milan in the summer of 1865 (Fig. 1). The new telescope was to be installed in the north wing of the Observatory, in the position previously occupied by the Amici reflector. The project for the new dome was started in 1863, but had to be interrupted due to budget limitations. Only in 1873 were new funds obtained from the government, allowing construction to begin. At the end of 1874, the dome was ready, and regular observations started in February 1875 (Fig. 2).

The telescope had a tube made of wood (fir) and brass and a German equatorial mounting with graduated circles at the end of both axes, and was equipped with a mechanical engine for the motion in right ascension. The construction of the clock drive was quite old-fashioned, since it lacked a pendulum regulator: the velocity

Fig. 1 Two images of the 22 cm Merz refractor: (*on the left*) a vintage photo (epoch unknown, but surely before WWII); (*on the right*) a contemporary photo, after the restoration by ARASS (June 2010)

Fig. 2 A recent image (March 2014) of the dome of the 22 cm Merz refractor

was limited by two small paddles rotating within a circular box and was controlled only by friction (Fig. 3). Actually, Schiaparelli often complained about the lack of accuracy of such a mechanism, which required frequent tuning. The base of the telescope was a pyramidal structure of cast iron and was built in Milan by the Tecnomasio firm; originally, the base was decorated and made heavier by cast-iron tiles, which were subsequently lost (as it appears by comparing the two pictures in Fig. 1). The objective-lens was of good quality, with a resolution approaching the theoretical limit for its diameter, and had a slight greenish coloring which Schiaparelli did not consider a flaw, since it added contrast in the observations of Mars. The equipment of the telescope included 13 eyepieces (allowing magnification factors from 67 to 690) and one filar micrometer (Miotto et al. 1989).

The new telescope was mainly used by Schiaparelli for observations of asteroids and comets, for micrometric measurements of double stars, and for the determination of the rotation periods of Mercury and Venus, but it was also used for his first observations of Mars, which started in an almost fortuitous way. On August 23rd, 1877, he was observing a total lunar eclipse and had no plans for measurements of double stars ("*with double * nothing can be done*" "*colle * doppie niente vi è da fare*"—he annotated in his logbook); he then aimed the telescope at Mars, which, at that time, was in a very favorable opposition. The rough sketch of the planet made during that night was the first step of his famous first essay on Martian rotation and topography (Schiaparelli 1878).

Fig. 3 The clock-drive of the 22 cm Merz telescope

1.2 The Large Merz Refractor

After the success of this first series of observations, Schiaparelli gained popularity and was in a position to ask for a second, larger telescope. He wished to turn again to the Merz Company, which already had a 49 cm doublet available, built as a spare copy for an order by the Strasburg Observatory, but the Parliament—mainly under pressure from Giuseppe Marcora (1841–1927), representative of the electoral college of Milan—asked him first to consider the possibility of having the lens built by an Italian firm. Since no Italian factory was capable of building the mount, it was ordered directly from Repsold in Hamburg. In reality, the only factory which applied for the assignment was the Filotecnica Salmoiraghi, which had never produced a lens of such a size. Schiaparelli signed a preliminary agreement with this firm in July 1878 but Salmoiraghi, in accepting the commission, was biting off more than he could chew, and, after several delays and requests for postponement, at the end of 1879, he was compelled to renounce the commitment. Schiaparelli reconsidered the Merz offer but, since the lens that was available had a strong greenish hue, he asked the factory to build a second doublet of the same size, agreeing to choose one of the two. The agreement was signed in February 1880; by April 1881, the new lens was ready and, a month later, Schiaparelli went to Munich (together with Giuseppe Lorenzoni, director of the Padua Observatory; see Zanini, in this volume) to test the lenses. Ultimately, they selected the second lens. All the parts of the telescope were shipped to Milan between the end of 1881 and the

Fig. 4 An image of the dome of the 49 cm Merz-Repsold on the southern facade of Brera Palace, seen from Brera Botanical Garden. The epoch of the photo in unknown, but is before 1921, since the small dome on the top of the west tower (*on the left*), where the Merz-Cavignato would have been located, is not present

beginning of 1882. Schiaparelli decided to place the new instrument at the top of the Southern facade of the Observatory, dismantling the octagonal tower built by Boscovich in 1764–65 which no longer complied with the needs of modern observational astronomy (Fig. 4). Again, the installation was delayed by problems with the construction of the dome, which was carried out by a company (owned by Edoardo Süffert, of Milan) that had no experience in the field. The large cylindrical structure was very heavy and problems arose regarding its movement; moreover, the opening mechanism was faulty and allowed rain to seep in. Finally, the telescope was installed and regular observations started in May 1886 (Miotto et al. 1989; Tucci 2000).

The telescope had a focal length of 698 cm (f/14.3). The mechanism for the diurnal motion was driven by a weight sliding inside the iron column of the basement (400 cm in height). The tube of the telescope was made of laminated iron (Fig. 5). With this instrument, Schiaparelli continued his series of observations in all fields of his interest, in particular, the measurement of double stars and the study of the surface of Mars, for which the Merz-Repsold was used starting from the fifth opposition observed by him (Schiaparelli 1897).

Fig. 5 A picture of the 49 cm Merz-Repsold refractor in its original dome in Brera (epoch unknown, but probably 1923 or few years later, since the telescope is equipped with the Zeiss spectrograph)

1.3 A Third Merz Telescope

The final cost of the 49 cm lens turned out to be lower than the estimated amount that had already been paid to Merz by the Brera Observatory. As a reimbursement, Merz supplied Schiaparelli with an additional, smaller lens, a 16 cm doublet with a focal length of 137 cm. Schiaparelli was particularly happy with this acquisition, since the higher luminosity of the lens (f/8.6) made it suitable for his observations of comets. The mounting was built by the *Società Veneta per Imprese e Costruzioni Pubbliche—Officina dell'Osservatorio di Padova* under the direction of Giuseppe Cavignato, but Schiaparelli could not find a suitable location for it. Only in 1921 did Luigi Gabba (1841–1916), provisionally in charge of the direction of the Observatory, build a small metallic dome for it on the top of the west tower of the southern facade, allowing the Merz-Cavignato to be put into use. It was equipped with a micrometer and used for observations of asteroids and comets (Fig. 6).

Fig. 6 A recent image (July 2010) of the 16 cm Merz-Cavignato after the restoration, inside its original dome

2 The Merz Telescopes After Schiaparelli

After Schiaparelli's retirement (Fig. 7)[3] in October 1900, the two main Merz telescopes were used intensively by the astronomers at Brera. This once again confirms the excellent quality of the optics of these telescopes, used for almost a century. It must be mentioned that the Merz refractors in Brera were not an exception: other Merz telescopes in Italy (see Poppi and Altamore, as well as Massone, in this volume), most of which were modified and modernized, continued to be used well into the first half of the twentieth century for scientific and educational purposes.

[3]The reasons for Schiaparelli's retirement are best explained by his own words, written in the last page of his observation logbook: *Once back in Milan* [after a period of vacation], *I planned to continue the observations at least until the end of 1900: but my health was no longer good enough to support such an effort and my eyes became more and more ill-suited. I must resign myself here to ending my observations with the equatorial* [telescopes] *which I began in February 1875 with the 8 inch and carried out until 1889, then continued with the 18 inch for 15 years up until 1900. Total number of double stars measurements, 11,775 (Tornato a Milano confidava di poter ripigliare le osservazioni almeno fino alla fine del 1900: ma la salute non era più sufficiente per reggere ad una simile fatica e l'occhio sempre più disadatto. Devo pertanto rassegnarmi a terminare qui le mie osservazioni equatoriali, cominciate nel Febbraio 1875 all'8 pollici e con esso condotte fino al 1889, indi proseguite al 18 pollici per 15 anni fino al 1900. Misure di stelle doppie in tutto 11,775).* The last (11776th) micrometric observation is followed by the remark: *My last observation, badly carried out, which persuades me of the inevitable need to discontinue these observations. Cecidere manus [= the hands fell] Oct. 29th 1900. (Ultima mia osservazione male riuscita, la quale mi persuade della necessità inevitabile di non continuare queste osservazioni. Cecidere manus. 29 Ott. 1900).*

Fig. 7 First page of the issue of the *Domenica del Corriere* (the weekly magazine of the newspaper *Corriere della Sera*) dated October 28th, 1900, depicting Schiaparelli with the 49 cm Merz-Repsold. The cover, illustrated by Achille Beltrame (1871–1945), was devoted to Schiaparelli on the occasion of his retirement

2.1 The Fate of the 22 cm Refractor

After about sixty years, the 22 cm refractor showed signs of progressive degradation in its mechanical parts: it was therefore decided that the tube and the mounting should be replaced completely, keeping only the original lens. The new mounting was built in the 1930s by Salmoiraghi[4] and installed in the same dome, replacing the original instrument; the old mounting by Merz was disassembled and stored in the room of the Starke meridian circle, in the west tower of the Observatory (formerly the belfry of the old church of Santa Maria in Brera, which had been dismantled in 1808). During World War II, the Merz-Salmoiraghi was temporarily moved to Merate, near Milan, in order to protect it from possible damage, and was reinstalled in its dome at Brera after the end of the war. However, in the following years, the instrument was used less and less frequently, due to the troublesome conditions at the Observatory after the war and the increasing difficulty of observation owing to the polluted environment in the center of the town. The last scientific utilization of the Merz-Salmoiraghi at Brera seems to have been a series of observations of stellar occultations and lunar eclipses performed by Edoardo Proverbio in the late 1950s in order to improve the determination of the Ephemeris Time and Lunar orbit (Proverbio 1958, 1960). In 1961, the tube of the telescope was temporarily installed on a concrete pillar built on Monte Conero, near Ancona, for observation of the total solar eclipse of February 14th, 1961 (Zagar 1961); it was then taken out of service.

The original Merz mounting was damaged during World War II in the bombing of Brera Palace that occurred on August 8th, 1943; it was restored so that it could be put on display at the *Mostra storica della scienza italiana*, held in Milan (in the building of the former Royal Palace) from November 1957 to January 1958. In order to remove it from the ruins where it had been lying since the time of the bombing, the column of the telescope had to be sawed into two parts, which were subsequently soldered together and strengthened with steel tie-rods (Fig. 1b). Later, the telescope was given on an extended loan to the *Museo Nazionale della Scienza e della Tecnologia "Leonardo da Vinci"* in Milan, where it was kept on display until 1997. In 1998, it was restored by the *Associazione per il Restauro degli Antichi Strumenti Scientifici* (ARASS) and again placed in its original dome, where it is now accessible to the public during guided tours organized by the *Museo Astronomico* hosted by the Brera Observatory. The restoration included the reinstallation of the original lens and has brought the instrument back to its full operational state. Being that the Merz telescope was reinstalled at Brera, the Salmoiraghi mounting has been given in exchange to the *Museo Nazionale della Scienza*, where it is still on display.

[4]Unfortunately, we were unable to find any document in the archives of the Brera Observatory related to this operation, so that the dating of the refurbishment is uncertain and based on indirect evidence and oral tradition; the situation is worsened by the fact that, usually, in scientific publications, the new Merz-Salmoiraghi refractor was simply referred to as the "Merz telescope". In general, the documentation available in the archives about the 22 cm Merz (including any documents related to its acquisition, construction and installation) is highly deficient; there certainly should have been a complete dossier about this instrument, but if there was, for some reason, it is no longer extant.

2.2 The End of the 49 cm Refractor

In 1909, the 49 cm Merz-Repsold refractor was equipped with a new wide-field filar micrometer (by Welharticky und Pachner, Vienna), and in 1923, with a Zeiss spectrograph; however, the degraded conditions of the sky in Milan put severe limitations on the use of the instrument. In 1925, the Brera Observatory obtained a branch observing site in Merate, 30 km to the northeast of Milan, in a villa which had formerly been used as a hospital. Merate had been conceived from the beginning as a facility devoted to astrophysics: Emilio Bianchi (1875–1941), director of the Brera Observatory, built a large dome in the park of the villa, that hosted a new 102 cm reflector equipped with a spectrograph, both by Zeiss (Bianchi 1941). In 1936, the Merz-Repsold refractor was also moved to Merate, into a dome built expressly for the telescope, just in front of the Zeiss dome (Fig. 8); the new dome was equipped with a movable platform in order to facilitate the use of the telescope at different heights. On that occasion, the telescope was renovated, substituting the mechanical diurnal motion with a new electric engine; unfortunately, many parts of the original mechanics were lost in the process.

Fig. 8 The Merz-Repsold after the upgrade of 1936, in its new dome in Merate (date unknown, but probably shortly after the installation)

The old dome of the Merz-Repsold in Milan remained empty and unused, and was seriously damaged by the already-mentioned bombing of 1943, which completely destroyed the metal cylinder, leaving the masonry relatively unaffected. The dome was restored in 1957 when, in the framework of the International Geophysical Year, the Brera Observatory undertook an ambitious program in the field of astrometry and time determination: the dome was provided with a new covering consisting of four triangular slabs which opened outwards, like the petals of a flower, exposing two narrow stripes of sky in the north-south and east-west directions for the use of transit instruments. For this reason, the newly-built dome was immediately nicknamed *cupola a fiore (= flower dome)* (Fig. 9). The *cupola a fiore* was used for astrometric observations up to the beginning of the 1970s, then remained unused until 1995, when it was transformed into a lecture room. The architect Adalberto Caccia-Dominioni maintained the original covering, blocking it in a half-opened position and filling the resulting gaps with metal inserts.

In 1962, a new telescope was installed in Merate, with a reflector designed by Glauco De Mottoni (1901–1988) and built by the Ruths factory in Genoa; the telescope had a 137 cm primary mirror made of aluminum and, at that time, it was the largest aluminum mirror in the world (Broglia and D'Avanzo 2010). Since the Observatory could not afford construction of a new dome, the telescope was located

Fig. 9 A detail of the interior of the *cupola a fiore*, with two of the four hydraulic rams which opened the slabs of the covering, and one of the four passage instruments which were installed therein (probably around 1957)

in the dome of the Merz-Repsold, which was put out of use and disassembled. In order to take advantage in some way of the old 49 cm objective-lens, De Mottoni built a new tube for it by adopting a catadioptric design in which the optical path was folded by two flat mirrors. The aim was to obtain a shorter, easily maneuverable telescope which could be installed in a smaller dome. However, the result was far from satisfactory: in order to reduce the construction costs, the new instrument made use of the old equatorial mounting of the Merz-Repsold telescope, but, since the tube was much heavier (due to the additional mirrors), the telescope had serious problems in regard to stability and movement. Moreover, during the tests for adapting the mounting of the lens to the new tube, the large Merz objective-glass fell down and broke. In 1970, in a last attempt to complete the telescope, it was equipped with a new lens, a triplet having a diameter of 42 cm and the same focal length as the old Merz objective; the new objective was designed by C. Morais and built by *Officine Galileo* in Florence (Fig. 10). However, this telescope was never used: at the beginning of the 1980s, the Morais triplet was sold to the Turin Astronomical Observatory, and the instrument was dismantled.

In 2010, the Observatory undertook restoration of the original tube and mount of the Merz-Repsold, which had again been committed to ARASS. Presently, the restoration is almost complete, and the Observatory is looking for a suitable location for putting the instrument on display. This is not an easy task, since the two domes that have hosted the telescope in the past are no longer viable possibilities: the original dome in Brera

Fig. 10 A picture of the Morais-Galileo telescope inside its dome (in reality, a small rectangular building with a sliding roof, which is open in the photo). The epoch of the photo is probably 1970. It is apparent that the size of the counterweight is greatly enlarged with respect to Figs 5 and 8, in order to balance the increased weight of the tube

has been heavily modified into the *cupola a fiore* and can no longer contain the 700 cm tube, and the dome in Merate is occupied by the Ruths reflector.

2.3 The Merz-Cavignato

The dome and mounting of the Merz-Cavignato refractor were also damaged by the bombing of 1943. The objective-lens was spared because, during the war, as a precautionary measure, it was temporarily moved to Merate, like most of the lenses of the Brera instruments; it has been used for some years as an astrograph mounted on the Ruths reflector. In 2000, the Merz-Cavignato was also restored and has once again been placed in its original location.

Acknowledgements Although responsibility for this paper is entirely that of the author, he owes most of its content to two persons: Pasquale Tucci, who has had a major role in the conservation, cataloguing and promotion of the historical patrimony of the Brera Observatory; and Pietro Broglia who, having worked at the Observatory since the late 1950s, has directly witnessed most of the events of the past five decades, and has investigated its history with great curiosity. The author is also indebted to Agnese Mandrino for her support in browsing through the archives of the Observatory. I wish to thank Ileana Chinnici for her meticulous and careful revision of the preliminary text.

References

E. Bianchi, *La R. Specola di Merate e le sue ricerche*, Contributi del R. Osservatorio Astronomico di Milano-Merate, Nuova Serie **10** (1941).

P. Broglia, P. D'Avanzo, *Il "Ruths" di Merate: un pioneristico esempio di telescopio ottico dotato di specchio metallico*, Giornale di Astronomia **36/4**, 32–37 (2010).

E. Miotto, G. Tagliaferri, P. Tucci, *La strumentazione nella storia dell'Osservatorio Astronomico di Brera* (Unicopli, Milano, 1989).

E. Proverbio, *Occultazioni di stelle e loro riduzioni e osservazioni di eclissi negli anni 1956–1957*, Rend. Sc. Istituto Lombardo **A92**, 459–472 (1958).

E. Proverbio, *Osservazioni di occultazioni da parte della Luna e correzione del moto lunare*, Rend. Sc. Istituto Lombardo **A94**, 561–565 (1960).

G. V. Schiaparelli, *Osservazioni astronomiche e fisiche sull'asse di rotazione e sulla topografia del pianeta Marte fatte nella reale Specola di Brera in Milano coll'equatoriale di Merz durante l'opposizione del 1877* (Reale Accademia dei Lincei, Roma, 1878).

G. V. Schiaparelli, *Osservazioni astronomiche e fisiche sull'asse di rotazione e sulla topografia del pianeta Marte fatte nella reale Specola di Brera in Milano coll'equatoriale di Merz (opposizione del 1886). Memoria quinta* (Reale Accademia dei Lincei, Roma, 1897).

P. Tucci (ed.), *I cieli di Brera; astronomia da Tolomeo a Balla* (Università degli Studi di Milano, Milano, 2000).

F. Zagar, *L'eclisse totale di Sole del 15 Febbraio 1961*, Mem. Soc. Astronomica Italiana **32**, 365–372 (1961).

Merz Telescopes in Rome

Aldo Altamore, Francesco Poppi and Sabino Maffeo

1 Introduction

In the second half of the nineteenth century, astrophysical research was born in Rome, thanks to the early studies carried out by Angelo Secchi S. J. (1818–1878) and Lorenzo Respighi (1824–1889), who were, respectively, directors of the Collegio Romano and Campidoglio Observatories. For their pioneering research, both astronomers used Merz instruments, whose contribution was therefore crucial in the development of the "New Astronomy", as astrophysics was called at that time.

The two telescopes were indeed produced by the well-known company founded in Munich by Georg Merz (1793–1867) in collaboration with Joseph Mahler, and continued by his sons Ludwig and Sigmund (see Wolfschmidt and Kost in this volume).

These two instruments were smaller versions of the large achromatic telescopes placed at Pulkovo Observatory in 1839 and Harvard College Observatory in 1847, having 38-cm apertures, they were the greatest refractors in the world until 1865; when an Alvan Clark & Sons' 47-cm aperture refractor was installed at the Dearborn Observatory of the University of Chicago.

It must be mentioned that, between the end of the nineteenth and the first half of the twentieth centuries, two other large Merz telescopes—no longer extant—were operating in Rome: the 27-cm aperture refractor installed at Villa Cecchina, on the

A. Altamore (✉) · F. Poppi
INAF-Osservatorio Astronomico di Roma, Roma, Italy
e-mail: aldo.altamore@uniroma3.it

F. Poppi
e-mail: francesco.poppi@gmail.com

A. Altamore · S. Maffeo
Vatican Observatory, Castel Gandolfo, Italy

I. Chinnici (ed.), *Merz Telescopes*, Historical & Cultural Astronomy,
DOI 10.1007/978-3-319-41486-7_6

slope of Gianicolo hill, where Gaspare Stanislao Ferrari S. J. (1834–1903) built a private observatory in 1880, after the confiscation of the Collegio Romano (Udìas 2003, p. 194), and the Merz-Gautier 40-cm aperture refractor installed in 1908 in the massive Leonine tower, near the summer residence of Leo XIII (1810–1903), in the walls of Vatican City, after the establishment of the Vatican Observatory (Maffeo 2001).[1] The present work is mainly focused on the Merz telescopes of the Collegio Romano and Campidoglio Observatories, as they made major contributions to the development of astrophysics, while the small Merz telescopes currently belonging to the Vatican Observatory are described in the final section.

2 The Merz Telescope at the Collegio Romano Observatory

In the second half of the eighteenth century, Ruggero Boscovich S. J. (1711–1787) first had the idea of erecting an observatory above the church of St. Ignatius, which is part of the architectural complex of the Collegio Romano. The Observatory was to be erected in the area that had been destined for the church's dome, which was never built and was artistically simulated from the inside by the amazing perspective painted on the ceiling by Andrea Pozzo (1642–1709).

However, Boscovich's project was abandoned and the Observatory was built in what is today called the Calandrelli Tower, which housed the astronomical equipment until the second half of the nineteenth century.

In 1852, Angelo Secchi revived the idea of building an astronomical observatory on the roof of the church, with the notion that the four pillars that were supposed to hold up the dome could be a very stable support for the installation of the telescopes.

Taking advantage of the support of Pius IX, the *New Observatory* was built in a very short time by the architect Angelo Vescovali (1826–1895), and a Merz achromatic refractor (24.4-cm aperture, 433-cm focal length) was purchased (Fig. 1), thanks to the resources inherited and made available by Paolo Rosa S. J (1825–1874), one of Secchi's assistants, as well as to a highly favorable offer from the manufacturer Georg Merz for an instrument that, at that time, would have been considered a medium-high class instrument.

The refractor was installed on October 25th, 1854, in the large dome of the new Observatory and remained in operation until 1889, when it was replaced by a new Steinheil-Cavignato refractor (40-cm aperture, 500-cm focal length) (Tacchini 1901). The new director, Pietro Tacchini (1838–1905), decided to include Secchi's historical Merz telescope in the collection of the Museo Astronomico e

[1]The Vatican Observatory was moved to the site of Castelgandolfo in 1935; its headquarters (where many historical materials are on display) were relocated to Albano Laziale in 2009.

Fig. 1 The Merz telescope used by Angelo Secchi at the Collegio Romano Observatory, (From: Cacchiatelli and Cleter [1860–69])

Copernicano, which had been established in 1886 by the Polish historian Arturo Wolinsky and Tacchini himself (Calisi 2000).

The Merz telescope remained on display in the Museum until the 1950s (Fig. 2), when the director of the Rome Observatory, Giuseppe Armellini (1887–1958), decided to put it to work again, combining it with the Steinheil-Cavignato telescope that was operating in the main dome of Villa Mellini, the new site of the Rome Astronomical Observatory.

This was a fatal decision for the destiny of the instrument. On the night between July 15th and 16th, 1958, the combined Steinheil-Cavignato and Merz telescopes were destroyed by a fire. Because of the stress provoked by this disastrous event, Armellini died of a heart attack the day after.

Fig. 2 The Collegio Romano Merz telescope on display at the Museo Astronomico e Copernicano in the premises of Villa Mellini in the 1930s (INAF-OAR Archive)

The original premises of the Collegio Romano Observatory above St. Ignatius Church were essentially abandoned and subsequently degraded (Fig. 3) after the move to Villa Mellini, where the Museo Astronomico e Copernicano was also relocated.

Even though the Collegio Romano Merz telescope is no longer extant today, a twin instrument—though later modified—is still preserved. In 1865, a Merz telescope that was, in effect, identical to the Roman one was installed at the Palermo Astronomical Observatory (Chinnici and Brenni 2015). With this instrument, in 1871, Pietro Tacchini[2] began a collaboration with Angelo Secchi in order to compare solar observations, carried out with the two Merz telescopes, and establish a scientific society devoted to the study of solar physics (Chinnici 2008). The Palermo Merz telescope was "modernized" in the 1950s, undergoing irreversible modifications, and used for educational purposes until the 1970s; after a careful restoration (Brenni et al. 2001), it is now on display (Fig. 4) as part of the heritage of the Palermo Astronomical Observatory .

[2]Tacchini was astronomer at the Palermo Observatory before succeeding Secchi as director of the Collegio Romano Observatory in 1879.

Fig. 3 The granite pillar of the Collegio Romano today with Merz telescope today (*credit* Renzo Lay)

Fig. 4 The Palermo Merz telescope today (Museo della Specola, INAF-OAPa)

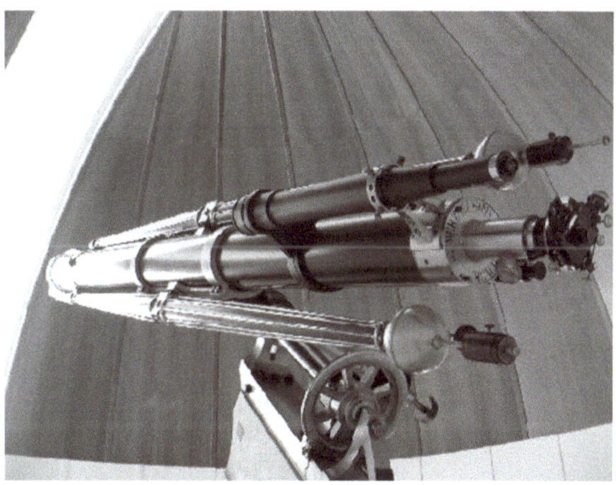

3 The Merz Achromatic Telescope at the Campidoglio Observatory

This telescope (11.4–cm [9”] aperture, 190–cm focal length) was donated to the University Observatory of Campidoglio in 1860, when it was under the directorship of Ignazio Calandrelli (1792–1866).

The donor was the Marquis Giuseppe Ferrajoli (1798–1870), an important businessman and politician in Rome during the years of Pius IX’s papacy.

It was inaugurated on the occasion of the solar eclipse that occurred on July 18th, 1860 (Scarpellini 1860) and was observed by Caterina Scarpellini (1808–1873), niece of Feliciano Scarpellini (1762–1824), founder and first director of the Observatory.

A few years later, it was used by Lorenzo Respighi to carry out his early studies on stellar astrophysics; the telescope was equipped with an objective prism (Chinnici in this volume) which he designed, based on an original idea by Joseph

Fig. 5 The Campidoglio Merz installed in the OAR station of Imperia (Italy) for observing the total solar eclipse of February 15th, 1861(INAF-OAR Archive)

Fig. 6 The Campidoglio Merz telescope, now on display at INAF-OAR (*credit* Marco Faccini)

Fraunhofer. The glass prism was provided by Merz to the Ertel factory in Munich, which made the mounting. Respighi also used the Merz telescope for solar observations and, by applying a wide slit spectroscope, he monitored the variations in solar prominences spanning several years (Altamore 2011).

In 1938, after the fusion of the Campidoglio and Collegio Romano Observatories, the telescope was moved to the eastern dome of the new Rome Astronomical Observatory (today INAF-OAR), located in Villa Mellini at Monte Mario, and remained in operation until the 1970s (Fig. 5). Today, it is included in the INAF-OAR historical collection and is on display at the Rome Observatory (Fig. 6), now located at Monte Porzio Catone.

The marble pillar, which originally supported the instrument at the Campidoglio Observatory, has recently been identified in the garden of Villa Mellini at Monte Mario (Fig. 7), where the headquarters of the National Institute for Astrophysics (INAF) are today installed.

Fig. 7 The marble pillar of the Campidoglio Merz telescope in the garden of Villa Mellini (*credit* Aldo Altamore)

4 Merz Telescopes at the Vatican Observatory

The third astronomical observatory in Rome, established at the end of the nineteenth century, was the Vatican Observatory—also known as Specola Vaticana (Maffeo 2001)—which took part in the international enterprise of *Carte du Ciel* (Chinnici 1999). The Observatory was originally located in the Tower of Winds, but the telescopes acquired over the years were installed in some existing towers along the Vatican walls. In order to equip the new Observatory, the barnabite Francesco Denza (1834–1894), its first director, acquired two Merz telescopes from the private observatory of Marquis Raimondo Montecuccoli Laderchi (1802–1873), which had been dismantled by his heirs. It is possible that a meridian circle, signed *K. K. polytechn. Institut in Wien. Chr. & G. Starke 1862*, once belonging to the Montecuccoli Observatory and acquired by the Specola around the same time, also had optical parts made by Merz, as other instruments built by Starke did (Zanini in this volume). In the collection of historical instruments kept at the Specola Vaticana, the two ex-Montecuccoli's Merz refractors are still preserved, together

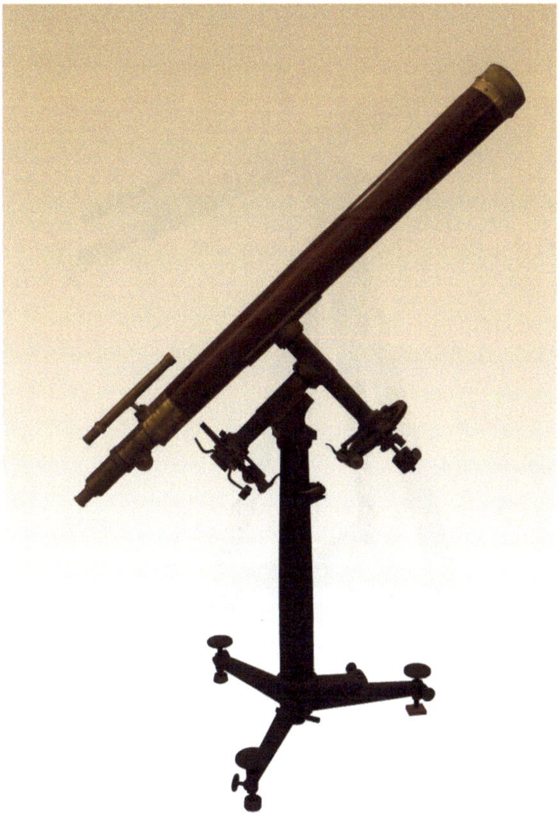

Fig. 8 The small equatorial telescope by Merz acquired from Montecuccoli heirs (*credit* Specola Vaticana)

with a third Merz telescope, purchased by Fr. Ferrari—former assistant of Fr. Secchi—for the already-mentioned private observatory at Villa Cecchina (Maffeo 2001, p. 24). A short description of the three Merz telescopes is given below.

4.1 Merz Telescope I

This is one of the telescopes belonged to the Marquis Montecuccoli (Fig. 8). It has an equatorial mounting, an aperture of 10 cm and a focal length of 1.5 m, and it is signed *G. Merz & Söhne in München*. It was located in the small dome on the half-tower near the grotto of Lourdes in the Vatican. Fr. Johann Georg Hagen S. J. (1847–1930), director of the Vatican Observatory from 1906 until 1930, used it to reexamine all the stars whose colors had been listed by Fr. Benedetto Sestini S.

Fig. 9 The altazimuthal Merz telescope of the Specola Vaticana (*credit* Specola Vaticana)

J. (1816–1890) at the Collegio Romano Observatory during the years 1844–1846, with the intention of publishing Sestini's list (Maffeo 2001, p. 375).

4.2 Merz Telescope II

This telescope (Fig. 9) was also acquired from Montecuccoli's heirs. It has an altazimuth mounting, an aperture of 10.7 cm and a focal length of 1.95 cm. It bears the same signature as the other telescope, *G. Merz & Söhne in München*, and can be dated to around 1850.[3] It was located in the half-tower right under the dome of the Merz Telescope I. It was mounted on a rail such that it could be pointed at the extreme eastern and western ends of the wall, so that it could be used to observe

[3]A similar Merz telescope, belonging to the private observatory built in Palermo by Prince Giulio Tomasi of Lampedusa (1813–1883) and acquired in 1852, is today kept at the Palermo Observatory (Chinnici and Randazzo 2011). In the collections of the Palermo Observatory, there is also a small telescope signed *Merz, Utzschneider et Fraunhofer à Munich* mainly used as comet seeker (Foderà-Serio and Chinnici 1997, p. 68).

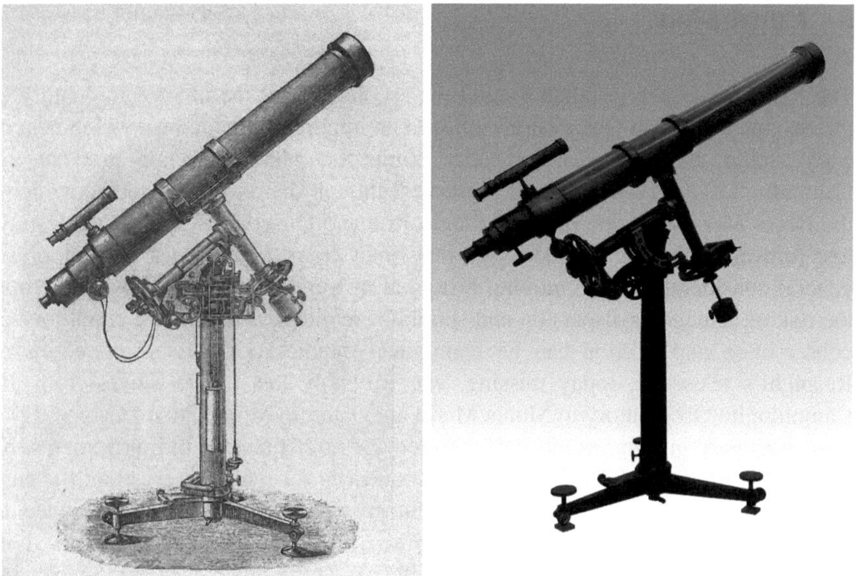

Fig. 10 *On the left* the small Merz equatorial telescope acquired by Fr. Ferrari in an engraving from Müller (1904, p. 191) and *on the right* as it is today (*credit* Specola Vaticana)

comets near the Sun during dusk and dawn. In 1995, it was given on temporary loan for a few years to the Discovery Park at Stafford, in Arizona (Maffeo 2001, p. 375); it is now on display at the Specola.

4.3 *Merz Telescope III*

This telescope dates back to 1880 and has an equatorial mounting, an aperture of 10.8 cm and a focal length of 1.5 m (Fig. 10). It has a wooden tube with a mahogany finish and is signed *G. Merz & Söhne in München.* The telescope was acquired by Ferrari for his private observatory at Villa Cecchina (Ferrari 1882; Müller 1904); it was sent to Russia with the Vatican expedition for the unsuccessful observation of the total solar eclipse of 1887. After the closure of the Cecchina Observatory in 1913, it became property of the physics laboratory at the Pontifical Gregorian University, upon which the observatory depended. Once courses ceased to be taught at the high school level, it was given back to the Specola and then got on loan by the Jesuit Leonine College at Anagni, near Frosinone. When the College transitioned to direction under the secular clergy, the telescope returned to the hands of the Jesuit Community at the Vatican Observatory.

5 Conclusions

The Merz telescopes installed around the second half of the nineteenth century in the two major Roman Observatories played an important role in the development of early stellar and solar astrophysics. While the Merz telescope used at the Campidoglio Observatory is still preserved, though deprived of its objective-lens, the larger Merz telescope of the Collegio Romano Observatory had a sad destiny, lost forever in the fire of 1958. We may infer from these stories two important general considerations: (1) moving historical instruments means exposing them to the risk of damage or dispersion and should be avoided, if reasonable conditions of conservation and fruition can be maintained on the spot: the objective-lens of Respighi's telescope, today missing, was probably lost in the moves from the Campidoglio Observatory to Monte Mario and, later, to Monte Porzio Catone; (2) it was customary in Italy, in the 1950s, to recycle or cannibalize old instruments, by dismembering them, with disastrous consequences for their preservation: the case of Secchi's Merz telescope is one of the most striking, but many other nineteenth century instruments, in many observatories, have been recycled, dispersed or irreversibly modified, thus rendering it difficult—sometimes even impossible—to recover the instrument. For example, the eyepiece holder of Secchi's Merz telescope is now mounted at the focus of the refractor hosted in the western dome of Villa Mellini. Only a few accessories, part of the right ascension circle, several eyepieces and a filar micrometer are properly kept in the collections of the Museo Astronomico e Copernicano, also located at Villa Mellini today, while the Merz objective prism used by Secchi (Chinnici in this volume) is on display at INAF-OAR of Monte Porzio Catone (Altamore et al. 2016).

In recent years, action towards preservation and fruition of the heritage related to the birth of astrophysics has begun[4] at INAF-OAR, both through the organizing of exhibitions and public events at Monte Porzio Catone, and through an ongoing project of requalification and restoration of the Museo Astronomico e Copernicano at Villa Mellini, which is supposed to be reopened to the public in autumn 2016. What remains of the two Roman Merz telescopes will finally find a proper location; their memory will not be lost.

The small collection of Merz telescopes at the Vatican Observatory had a secondary scientific role, but a remarkable historical significance, because they came from private observatories that are no longer extant and are probably the only instruments from these two small astronomical institutions that survived. They witnessed the contribution of the amateur Italian astronomers to the practice of astronomy, as well as their lost heritage, since only a few items from their equipment have been preserved. They have usually been acquired and used by institutional observatories, applying the above-mentioned policy of recycling and reusing old instruments, due to the lack of adequate financial resources. After restoration,

[4]A project of requalification of the now crumbling premises of the Collegio Romano Observatory is currently in preparation and is being promoted by the Jesuits overseeing St. Ignatius Church.

these telescopes are today preserved at the Specola, where they represent the history of astronomy at the Vatican and the contribution of the private observatories; they are, however, intended to be put on display in the Specola astronomical museum, whose realization is currently in progress.

References

Altamore A. 2011, in Altamore and Maffeo eds., *Angelo Secchi- L'Avventura scientifica del Collegio Romano*, Quater Edizioni, Foligno, p. 109.

Altamore A., Poppi F, Faccini M. 2016, in Chinnici ed., *Starlight – La Nascita dell'Astrofisica in Italia*, Arte'm, Naples, p. 48.

Brenni P., Chinnici I., Foderà Serio G., "The restoration of three large telescopes of the Palermo Astronomical Observatory", *Bulletin of the Scientific Instrument Society*, 71, December 2001, pp. 11–16.

Cacchiatelli P., Cleter G. *Le scienze e le arti sotto il pontificato di Pio IX*, Rome, [1860–69].

Chinnici I. 1999, La Carte du Ciel. Correspondance inédite conservée dans les archives de l'Observatoire de Paris, Osservatorio Astronomico di Palermo, Palermo.

Chinnici I., "The Società degli Spettroscopisti Italiani: birth and evolution", *Annals of Science*, vol. 65, No. 3 (July 2008), pp. 393–438.

Chinnici I. and P. Brenni, "The Palermo Merz Equatorial Telescope: an instrument, a manuscript, some drawings", *Nuncius* 30 (2015), 228–279.

Chinnici I. and D. Randazzo, "Old Astronomical Instruments on a Movie Set: the case of 'The Leopard'", *Bulletin of the Scientific Instrument Society*, 109, June 2011, pp. 9–12.

Calisi M. 2000, *Storia e Strumenti del Museo Astronomico e Copernicano di Roma*, OAR-INAF, Rome.

Ferrari G. 1882, Il nuovo osservatorio astronomico privato sul Gianicolo, Pontificia Università Gregoriana - continuation of the Bullettino Meteorologico dell'Osservatorio del Collegio Romano, XXI, Nos. 1 and 4.

Foderà Serio G. and I. Chinnici 1997, *L'Osservatorio Astronomico di Palermo*, Flaccovio Editore, Palermo.

Tacchini P. 1901, Mem. R. Oss. Coll. Romano, Ser. III, v. I.

Scarpellini C. 1860, Bull. Corrisp. Scient. di Roma per l'avanzamento delle Scienze, anno XII, n. 25.

Maffeo S. 2001, *The Vatican Observatory. In the service of nine Popes*, Vatican Observatory Publications, Rome.

Moigno F. 1879, *Le révérend Père Secchi. Sa vie*, Gauthier-Villars, Paris.

Müller A. 1904, *Elementi di astronomia ad uso delle scuole e per istruzione privata*, Desclee, Lefebvre E. C. Editori, Roma.

Udìas A. 2003, Searching the Heavens and the Earth: the History of Jesuit Observatories, Kluwer Academic Publishers, Printforce, The Netherlands.

On the Collection of Merz Instruments at the Naples Observatory

Mauro Gargano

1 The Collection

Over two hundred years of observations, the Astronomical Observatory of Capodimonte has acquired many instruments manufactured in Munich, firstly by the Reichenbach & Utzschneider Institute, where the young optician Joseph von Fraunhofer (1787–1826) carried out his famous experiments on the solar spectrum and made powerful objectives and oculars lenses, and later by the Georg and Sigmund Merz, successors to the company housed in the ancient monastery at Benediktbeuern.The largest telescope made in the first two decades of the nineteenth century was constructed by Reichenbach and Fraunhofer in 1814. It was commissioned by the Neapolitan astronomer, Federigo Zuccari (1783–1817), for the new astronomical observatory built on the Capodimonte hill. This equatorial telescope (17.5-cm aperture and 302-cm focal length)[1] was, at the time, "*the biggest one in the world, a real masterpiece*", as attested to by the astronomer Franz Xaver von Zach (1754–1832) (Zach 1819). The optical parts of the telescope, especially the objective-lens, were defined as being "excellent" by Carlo Brioschi (1781–1833), the first director of the Capodimonte Observatory (Brioschi 1824–26). It is possible that Georg Merz (1793–1867) was studying with Fraunhofer while he polished the lenses for this telescope and built the other instruments that the King of Naples, Joachim Murat (1767–1815), had ordered to purchase:[2] two repeating

[1]For a complete description of this telescope, as well as the other instruments mentioned here, see *Polvere di Stelle*, the web portal for the cultural heritage of Italian astronomy: www.beniculturali.inaf.it.
[2]About the delivery of these instruments and the visit of Reichenbach and Zach to Naples, see Gargano et al. (2015).

M. Gargano (✉)
INAF-Osservatorio Astronomico di Capodimonte, Naples, Italy
e-mail: mauro.gargano@inaf.it

I. Chinnici (ed.), *Merz Telescopes*, Historical & Cultural Astronomy,
DOI 10.1007/978-3-319-41486-7_7

circles, a transit instrument, a meridian circle, a good pendulum clock, and the equatorial telescope (ϕ = 8.3 cm, f = 120 cm), later used by Annibale De Gasparis (1819–1892) to discover nine asteroids.

In the second half of the nineteenth century, when Georg Merz continued the production of optical instruments after Fraunhofer and provided all of the main European observatories with his excellent telescopes, the Naples Observatory acquired some instruments to replace its obsolete equipment: in 1863, a new equatorial telescope, provided with an annular micrometer; in 1870, a large meridian circle made by Repsold with an objective-glass manufactured by Merz, and a spectroscope, by Merz as well, purchased by De Gasparis in the same year.

Finally, in 2012, on the occasion of the bicentennial celebrations of the Capodimonte Observatory foundation, in the midst of the setting up of the new outfitting for the MuSA-Museo degli Strumenti Astronomici (Museum of Astronomical Instruments), an Ertel universal theodolite, equipped with lenses by Merz, and a 11-cm aperture equatorial telescope, made by Merz in the second half of the nineteenth century were retrieved, about one hundred and fifty years after their acquisition.

2 The First Telescope Acquired by the Italian Government

Since the laying of the foundation stone, placed on November 4th, 1812, at a solemn ceremony, the Naples Observatory was equipped with an important collection of instruments, mainly used for celestial mechanics and topography studies, *to place the nascent Royal Observatory in Naples among the most notable of Europe with honor [...] for the progress of so a beautiful science as astronomy* (Visconti 1832).

The first observation was made on the night of December 17th, 1819, by Brioschi, who observed α Cassiopeiæ with the repeating circle housed in the eastern dome of the Observatory (Gargano 2016). Later, in 1834, the director Ernesto Capocci (1798–1864) had the repeating circle dismantled so as to replace it with the large Fraunhofer telescope. The year 1849 was a breakthrough year for the Naples Observatory, and a turbulent one: on one hand, De Gasparis discovered Igea Borbonica, the first of the list of nine asteroids discovered by him, but, on the other hand, Capocci was removed from the Observatory because of his liberal political ideas. The renewal of the instrumentation aimed at new scientific challenges seemed to have suffered a setback. While the political events of 1815 had affected the completion of the Observatory's construction,[3] contrastingly, the year 1860 marked a "new beginning" for the Capodimonte Observatory.

[3]In that same year, Joachim Murat was driven from the throne of Naples, the Kingdom's capital, and shot on October 15th in Pizzo Calabro, and King Ferdinand of Bourbon returned to Naples.

The Expedition of the Thousand, led by Giuseppe Garibaldi (1807–1882) to conquer the Kingdom of the Two Sicilies, ended in Naples, on September 7th, with the triumphant entry of the Hero of the Two Worlds. One week later, Garibaldi reinstated Capocci as director of the Astronomical Observatory. Among his first acts, the astronomer wrote a report to the Ministry of Public Instruction on the state of the Institute. Paraphrasing the words of his uncle Federigo Zuccari, who described the conditions of the astronomical instruments in 1811 by noting that *in the hands of a blind man, they were a free-for-all*, Capocci wrote in his report:

> With regard to the scientific conditions, not only is our Astronomical Observatory at the level of the observatories of the most remarkable capitals of Europe, but it also cannot be compared with those of the other provinces of Italy. To provide it with instruments, telescopes and other equipment suitable for the current advancements in science, it would require at least [...] 170,000 Lire, conveniently. But if you believe that purchases must be as thoroughly restricted as necessary, [...] 51,000 Lire would be enough for now.[4] ("*Stato* [...] 1861)

Some days later, on February 15th, 1861, Capocci embarked from Naples to Genoa to attend the first session of the Italian Parliament. Along with De Gasparis, he was a Senator of the Kingdom for *the eminent merit wherewith he illustrated the Fatherland*. The journey of the Neapolitan parliamentary delegation was very turbulent:

> Finally, we are in Turin after a horrible sea voyage [...] Our National Guards have suffered horribly, and cry out for revenge [...] They did not have but hardtacks and cheese, they were stowed below deck during a horrendous rain, it seemed all hell broke loose. Now everything is over, and they are amusing.[5] (De Sanctis 1861)

The author of this letter was Francesco De Sanctis (1817–1883), an intellectual who shared the same political views as Capocci. On March 17th, 1861, De Sanctis was appointed Minister of Public Instruction, and a few days later, the Neapolitan astronomer wrote him a heartfelt letter *to plead, his love for the sciences and national decorum being well-known, for some actions that are now required because of the time elapsed*. After describing the degraded state of the Observatory and stressing its importance, due to the favorable location and climate conditions, hc concluded:

---

[4]*In quanto poi alle condizioni scientifiche, la nostra Specola Astronomica ... non solo non è a livello degli Osservatori delle più cospicue Capitali di Europa, ma non può stare a confronto neppure con quelli delle altre province d'Italia. Per provvederla convenientemente di macchine, di cannocchiali e degli altri strumenti richiesti dagli odierni progressi della scienza, occorrerebbero almeno Duc. 40,000 pari a lire 170,000. ma laddove si credesse restringere gli acquisti al puro necessario,basterebbero per ora Duc. 12,000 pari a lire 51,000.*

[5]*Dopo un orribile viaggio di mare eccoci alfine a Torino ... Le nostre guardie nazionali hanno orribilmente sofferto, e gridano vendetta ... Non hanno avuto che gallette e cacio, e sono rimasti stivati sopra coperta, sotto una pioggia orrenda che parea il finimondo. Ora tutto `e finito e se la passano allegramente.*

> I beg you now to provide this Observatory, without further delay, with an achromatic telescope having an objective from five to six inches diameter, mounted in a special way and also equipped with micrometers, to be used for searching for new planets and comets, and determining their positions accurately by means of differential observations.
>
> And in order to better succeed in this aim, I would propose that you send Prof. De Gasparis to London to select the one to be purchased; he will be the one to use it most in the sort of researches for which he has achieved so much celebrity.[6] (Capocci 1861a)

The general secretary of De Sanctis, Quintino Sella (1827–1884), informed Paolo Emilio Imbriani (1808–1877), the representative of the Minister of Public Instruction in Naples, *to give special consideration to the request of the R[oyal] Observatory of Naples* (Sella 1861); in spite of an initially positive reply from Imbriani, nothing happened, and Capocci's request was renewed in December 1861. He wrote two private letters to De Sanctis and Francesco Brioschi (1824–1897), a mathematician and general secretary of the Ministry of Public Instruction, *to recommend this important institute* (Capocci 1861b); he addressed the Minister with the hope that *Italy will not refuse to do in some small part what Great Britain, France, Russia,* etc., *have done for their observatories in far greater proportions,* and added, *[the Naples Observatory] deserves the special favor of the government more than any other in our vast Kingdom due to the serenity of the weather of the site, and specifically the strength and magnificence with which it was built; and a Neapolitan will not fail to distinguish his administration by financing it.*[7] (Capocci 1861c)

The other astronomer-Senator from Naples, De Gasparis, facilitated Capocci's request. He wrote a letter to Sella to request new instruments for the development of Astronomy, as came to be written in words over the Observatory's main entrance:[8]

> Because we are lacking in good instruments, it is impossible for me to search for the planet of Schiaparelli.[9] Because of the smallness of this celestial body, the very few times that it

[6]... *io mi fo a supplicarla di fornire per ora questo Osservatorio, senz'altro indugio, di un cannochiale acromatico di cinque in sei pollici di apertura, montato in guisa speciale, per servire alla ricerca di novelli pianeti e comete, corredato eziandio de' micrometri relativi, per accuratamente determinare le loro posizioni mediante osservazioni differenziali. E per vie meglio riuscire nell'intento io proporrei d'inviare a Londra il Sig.r Prof. De Gasparis a farne la scelta e l'acquisto; dovendo egli principalmente adoperarlo in siffatte ricerche nelle quali egli si è acquistato tanta celebrità.*

[7]*L'Italia non ricuserà di fare moderatamente ciò che in più ampie proporzioni fanno pe' loro osservatori l'Inghilterra, la Francia, la Russia ecc. Il nostro poi, per la serenità del clima ov'è posto, e per la solidità e magnificenza con cui appositamente fu edificato, merita il particolare favore del Governo più che qualunque altro nel nostro vasto regno; ed un napoletano non mancherà d'illustrare la sua amministrazione col procurarglielo.*

[8]In April 1819, once the work on the new building of the Naples Observatory had been completed, Giuseppe Piazzi (1746–1826), General Director of the Naples and Palermo Observatories and former Director of the Astronomical Observatory of Palermo, inserted into the gable of the building the inscription: ASTRONOMIÆ INCREMENTO.

[9]Giovanni Virginio Schiaparelli discovered the asteroid Esperia on April 29th, 1861, two months after the discovery of Ausonia, the asteroid found by De Gasparis on February 10th.

will remain visible, [...] the observations made at other Italian sites would be valuable for calculating its orbit.

As far as I know, only Father Secchi, who has a magnificent equatorial Merz, has observed it in Rome. I say this in writing, not because I think you need arguments to become interested in the idea of our science thriving.[10] (De Gasparis 1861)

In the initial months of 1862, Capocci obtained the requested funds, probably from the new Minister, the physicist Carlo Matteucci (1811–1868),[11] who was in Naples on May 26th to visit the scientific and cultural institutions of the City: *He showed how much he cared about the progress of liberal arts, scientific and artistic institutions of the country; he gave benefits, and accorded subsidies* ("*Visita* ..." 1862). In June 1862, Capocci wrote:

The Observatory is located in a beautiful place, on a hill with a broader horizon, the building has all the solidity required by the needs of science, and holds several instruments ...

In addition the Observatory has commissioned from März [sic] of Munich another 5.8 inch refractor with an equatorial mount and a clockwork mechanism, the current Minister having granted £8000 for this purpose during his visit to Naples. This telescope (with its discrete size and being more powerful than all the currently existing equatorial telescopes) could be useful for observing new planets, comets, double stars, and so on. It can be placed in the retractable roof dome at the smallest cost.[12] (Capocci 1862a)

The acquisition of the first telescope purchased by the Italian government was therefore the result of the common effort of a class of politicians and scientists who were all fervent patriots, sharing the same political ideals of the Risorgimento and the same interest in the advancement of public education and scientific research.[13]

In July 1862, Georg & Sigmund Merz proposed to Capocci either a 24-cm aperture telescope, like the one at the Collegio Romano Observatory, or a 22-cm objective telescope, like the one sold to Domenico Ragona (1820–1892) for the

[10]*La mancanza in cui siamo di buoni strumenti mi ha reso impossibile la ricerca del pianeta di Schiaparelli. Attesa la piccolezza di questo astro, il pochissimo altro tempo che rimarrà visibile, [...] le osservazioni fatte da altri siti in Italia sarebbero riuscite preziose per il calcolo dell'orbita. Per quanto è la mia notizia, il solo padre secchi a Roma, provvisto di un magnifico equatoriale di Merz, lo ha osservato. Dico ciò per istesso, non perché credo che abbiate bisogno di argomenti onde interessarvi a far prosperare la nostra scienza.*

[11]Matteucci is well-known for his studies on magnetism and electrolysis and for being one of the founders of the journal *Il Cimento* in 1844.

[12]*L'osservatorio posto in bel clima, su una collina col pi ampio orizzonte, e costruito con tutta la solidità richiesta dai bisogni della scienza, possiede varî strumenti ... Oltre ai detti strumenti si è commesso a März di Monaco un altro rifrattore 5,8 poll. Montato parallatticamente con moto di orologeria, avendo per tal uopo accordato S.E. il Ministro presente 8000, nella sua venuta in Napoli. Questo cannocchiale (di discreta mole e di penetrazione molto maggiore dell'attuale macchina equatoriale) può riuscire utilissimo alle osservazioni de' nuovi pianeti, delle comete, delle stelle doppie ecc. Potrà collocarsi con lievissima spesa nella cupola con tetto mobile.*

[13]A few months later, a much more substantial sum, 19,400 Lire, was given to Giovanni Virginio Schiaparelli (1835–1910) to purchase the equatorial Merz telescope with a 22 cm objective used by the astronomer of Brera for his first world-famous observations of the planet Mars (Lombardi and Mandrino 2008) (see also Carpino in this volume).

Fig. 1 The polar axis of the telescope with its right ascension circle, the silver graduated limb, and a nonius. On the right, the wheels and the pinion of the Merz original clockwork movement (MuSA)

Palermo Observatory (Merz 1862a). Considering the available budget, the astronomer of Capodimonte commissioned a "*réfracteur de 60 lignes d'ouverture monté parallactiquement* (Merz 1862b). The first telescope purchased with funds from the new Kingdom of Italy was delivered to the Naples Observatory at the beginning of September 1863 (Capocci 1862b).[14]

The telescope has a 13.5-cm objective and a 220-cm focal distance and is signed G. & S. Merz in München. The equatorial circle has a silver graduated limb with a 22-cm diameter divided into 1'; it is equipped with two verniers allowing to measure hour angles within 4" (Fig. 1).

The declination circle, having a 28-cm diameter, is divided into 10' and has two verniers which enabled the user to read the declination angles with an accuracy of 10". The telescope was originally equipped with a conical pendulum driving-clock, a filar micrometer, some eyepieces and an annular micrometer.

On January 6th, 1864, Capocci died of an aneurysm, and Annibale De Gasparis was appointed as his successor. The new director of the observatory urged the talented mechanic, Pasquale Moreno (1859–1922), to complete the work in the eastern dome so as to be able to house the new telescope as soon as possible. By the

[14]However, the telescope was not immediately installed in the eastern dome, because Capocci retired to his native village, Picinisco, *around October 1863 when his health began to suffer* (Del Giudice 1865).

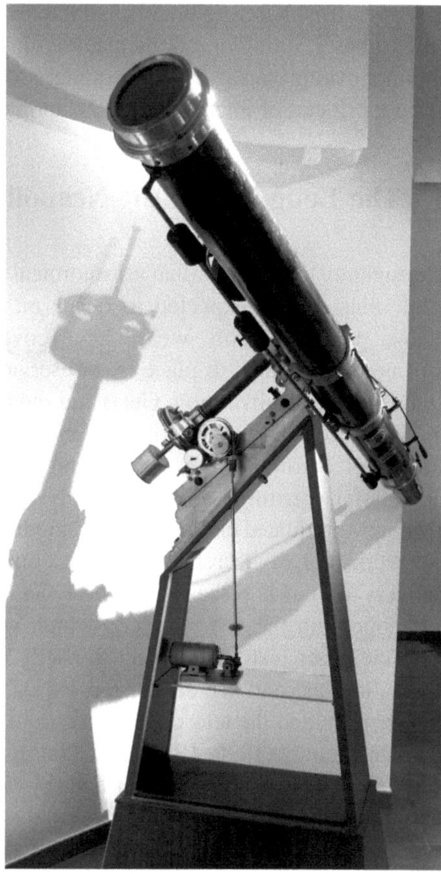

Fig. 2 The restored Merz telescope in the new exhibition at the Astronomical Museum of the Capodimonte Observatory; it is installed on a modern column designed after the original. In the foreground, the electric speed control engine and the rod transmitting the movement (MuSA)

beginning of 1865, all the dome fittings were completed and the Merz telescope was placed *on a wooden stand lying directly on the floor of the room.* A Le Roy astronomical pendulum clock, without compensation and with a retreating escapement, was also placed in the room (De Gasparis 1884). In 1875, the telescope was additionally provided with an old Gambey filar micrometer, bought in Paris around 1840 at the behest of Capocci, and modified by the Neapolitan instrument-maker Giuseppe Spano (1806–1873).

To test the telescope's quality, the Neapolitan astronomer Emanuele Fergola (1830–1915) measured the positions of six sample stars (α Tauri, ε Leporis, β Orionis, α Leporis, α Columbae, and μ Gemini). He found that the corrections for the hour and declination circles were $\Delta t = + 33\ ^{s}.8$ and $\Delta d = -5,$" and the flexure was $f = +6.1$. Consequently, the errors in Right Ascension and Declination measurements were about 8 and 0.5 times the respective accuracy of the graduated circles (Fergola 1865a, b). Due to these results, the expectations of Capodimonte astronomers with regard to the new telescope were partly disappointed: *despite the celebrity of the author, we have not found it to be up to the hype* (*Sulle condizioni*

... 1869). It is therefore understandable that, in the literature, there are no scientific articles based on observations made with this instrument until 1870, when a lucky chain of events caused the most modern telescope of the Capodimonte Observatory to be the focus of interest of the Neapolitan astronomers.

3 The Long Life of the Neapolitan Merz Telescope

Throughout 1869, the Italian astronomical community was greatly excited about the total solar eclipse expected to occur on December 22nd of the following year, whose path of totality would have covered eastern Sicily. A board of Italian astronomers decided to place two observational stations, one in the old Capuchins convent at Terranova (now Gela) and the other on the terrace of the Swabian castle in Augusta. The commission focused the scientific program on the spectroscopic studies of chromosphere, corona and solar prominences, and decided to use the Fraunhofer equatorial telescopes of Naples and Florence as the main instruments. Two new graduated circles for the Capodimonte telescope were commissioned from the workshop of Secrétan[15] in Paris, but the French-Prussian war prevented the delivery of those pieces. Then, De Gasparis decided to replace the Fraunhofer telescope with the one by Merz, which was shipped in November 1870, together with the other instruments sent to Sicily. Arminio Nobile (1838–1897)[16] was the Neapolitan astronomer designated by De Gasparis to take part in the expedition. Nobile provided the telescope with an old spectroscope, built in 1866 by Giovanni Battista Donati (1836–1873) and Giuseppe Poggiali (1820–1892) and equipped with an excellent prism furnished by Father Angelo Secchi (1818–1878). In Terranova, Nobile observed the coronal spectrum, measuring the Fraunhofer lines and also *the one that Kirchhoff and Angström ascribed to iron vapours*.[17] Nobile wrote in his report:

---

[15]August Charles Louis Secrétan (1833–1874) was the son of Marc François Louis (1804–1867), who was in partnership with Nöel Marie Paymal Lerebours (1807–1873), a famous opticians and daguerreotypist, in a scientific instrument workshop founded in 1845. Ten years later, Lerebours, one of the founders of the first photographic society, the Société Héliographique, left the workshop and retired to Neuilly-sur-Seine, just west of Paris. In 1867, the firm was carried on by August and his cousin George (1837–1906) under the name of "Secrétan, opticien de S. M. l'Empereur à Paris" (Brenni 1994).

[16]Arminio Nobile was the son of the astronomer Antonio (1794–1863) and the poetess Maria Giuseppina Guacci (1807–1848). He studied at the "Scuola di applicazione per ingegneri di ponti e strade", the ancient Neapolitan engineering school, and became a lieutenant in the Italian Army artillery. When his father died, Arminio Nobile became a junior scientist at the Observatory, and the following year, he was appointed as an assistant (Gargano 2013).

[17]Nobile also aimed the Merz telescope at *the cusps producted by the Moon on the Sun* before the totality phase of the eclipse, and observed some spectral lines, such as Hα, Hβ and NaI (Olostro Cirella and Gargano 2016). However, he was never able thoroughly to exploit enough this brilliant insight that crowned his ability as an excellent and hard-working sky watcher (Capaccioli and Gualano 2012).

> The spectrum of the corona presented me with a single shiny line in the green that probably corresponded to 1474 Kirchoff, as is also evident from the measures I took. This line belongs to the iron. I have not seen any other line, although the optical conditions of my instruments are good *[...]* It was impossible for me to identify the expected line of the northern lights. I must add that the bright line observed in the corona did not disappear when the solar limb reappeared, but gradually began to die out about 15 [seconds] after the end of the totality. I have no more to say about the spectrum of the solar corona.[18] (Nobile 1872)

Once back to Naples, the telescope was reinstalled in the eastern dome of the Observatory, and was used for some studies on double and multiple stars, almost exclusively by Nobile, who was the only one to have acquired the necessary expertise.[19] In 1892, the telescope was finally removed to be replaced by the Wanschaff zenith-telescope purchased by Fergola, who, 27 years earlier, had tested the *non excellent* quality of the Merz telescope. It was then put aside and was intended to be kept in a Observatory's warehouse, in some boxes. The telescope's "career", however, was far from being over. In 1919, the director of the Capodimonte Observatory, Azeglio Bemporad (1875–1945), made available the Merz telescope to Giovanbattista Lacchini (1884–1967), a Post officer and an amateur astronomer in Faenza (near Ravenna), for his research on variable stars. In December, Bemporad asked to the Minister of Education for the permission to loan it:

> Mr. G.B. Lacchini, [...] is a well-known and active member of the American Association of Variable Star Observers, and an author of valuable works [...] has asked to borrow the equatorial Merz telescope with a 15 cm aperture, very suitable for the kind of work in which Lacchini has acquired deserved fame. Here, the instrument is lying idle for lack of proper personnel [...] the Observatory is interested as well as Mr. Lacchini for the systematic observations of variable stars.[20] (Bemporad 1919)

---

[18] *Lo Spettro della Corona non ha presentato a me che una sola riga lucida nel verde che corrispondeva probabilmente alla 1474 Kirchoff, come risulta anche dalle misure prese. Questa appartiene al ferro. Non ho veduta nessuna altra riga, quantunque le condizioni ottiche dei miei strumenti fossero buone ... Non fu a me possibile discernere l'aspettata riga dell'aurora boreale. Debbo aggiungere che la riga brillante veduta nella Corona non spari al momento della ricomparizione del lembo solare, ma andò gradatamente spegnendosi durante circa 15 [seconds] dopo la fine della totalità. Altro non ho a dire sullo Spettro della Corona.*

[19] In 1877, Dom Pedro II, the Emperor of Brazil, during a visit to the Observatory, indicated that he wanted to point this telescope skyward from Naples to observe some multiple star systems, such as the open cluster Trapezium in the heart of the Orion Nebula (Bianchi 2012). The Florentine newspaper *La Nazione* reported the news of the imperial visit: *... as a man skilled in the astronomical art, he watched some multiple star systems with the Merz equatorial telescope [and] he talks about astronomy not only as a man who has studied it, but also as a scientist who is closely following its advances.* (*La Nazione*, February 8th, 1877)

[20] *Il sig. G.B. Lacchini [...] noto ed attivissimo membro della"American Association of Variable Star Observer", autore di pregiati lavori [...] ha chiesto in prestito l'equatoriale di Merz di 15 cm di apertura, adattatissimo per il genere di lavori nel quale il medesimo Lacchini ha acquistato meritata fama. Lo strumento [giace] qui inoperoso, per mancanza di personale adatto [...] [essendo l'Osservatorio] non meno interessato del Sig. Lacchini alla osservazione sistematica delle stelle variabili.*

Fig. 3 Giovanbattista Lacchini and the Merz equatorial telescope at the Urania Lamonia Observatory in Faenza, around 1922. It is recognizable the driving-clock mechanism and its gravity-driven weight, lodged very compactly in the wooden pedestal, close to the polar axis of the telescope. (AAVSO, Historical Archives)

The Minister approved, and Lacchini was very excited: *I have no words to thank you for the great result reached in my favor for obtaining the loan of the instrument* (Lacchini 1919). Three years later, in 1922, the telescope was moved to Lacchini's private observatory, called Urania Lamonia (Fig. 3). The telescope remained in Faenza for ten years, serving in an intense observing campaign of variable stars.[21] In 1934, the telescope returned to Capodimonte, without the great fanfare given for its tenure gained in Faenza. In the manner of a historical bookends, the "scientific life" of the Merz telescope, started with the solar eclipse of 1870, closed, in 1936, with the observation of a total eclipse of the Moon: Maria Viaro (1907–?), the first female astronomer of the Capodimonte Observatory, used it to measure *the instants of some stellar occultations*. Later, the telescope was housed in the north dome of the Observatory; it was equipped with a new electric clockwise motion and a wide-field astrophotography objective (maybe the "Big Bertha" realized by Dallmeyer around 1940 for the British Royal Air Force) and occasionally used for

[21]For the decade from 1923–1933, the observations carried out with the Neapolitan telescope by the amateur astronomers in Faenza may be summarized by the statistics published in the magazine *Popular Astronomy*: 10,022 observations, of which over 6000 were made by Lacchini, of about 200 variable stars.

some studies of stellar photometry and transient objects (Perrotta and Gargano 2016). In 2008 the telescope was moved from the northern Observatory dome and it was completely restored. Then it was placed on a modern column similar to the original masonry one and exhibited in the museum (Fig. 2).

From 1864, when the Merz equatorial telescope was installed in the eastern dome of the Capodimonte Observatory, until 2008, when the instrument was removed from the northern dome, restored and reassembled in the museum, the telescope has undergone a back-to-back modifications and renovations: therefore, some parts were written off or lost, such as the original clockwork mechanism.

In the library and archives of the Capodimonte Observatory, there are no historical documents, sketches or drawings illustrating the telescope and its accessories. Surprisingly, some time ago, a photograph of the Merz telescope had been found in the AAVSO photographic archives: it shows the telescope when it was installed in Lacchini's private observatory in Faenza. The photo was probably taken in the first months of 1922, after the telescope's installation was completed.From an article by Nobile, we learn additional information about the clockwork and its workings:

> The clockwork apparatus with conical pendulum. To regulate the speed, we have to lengthen or shorten the small pendulum. In addition to the weights provided by the manufacturer, we usefully added more others, since the instrument is subject to stopping.[22] (Nobile 1875)

The correspondence between the amateur astronomer of Faenza and the director of the Capodimonte Observatory shows that, when the telescope was given back in 1934, the finderscope and the clockwork mechanism, described by Lacchini as being defective because of its irregular movement, were not sent to Naples.

As concerns the filar micrometer of the telescope, we only know that it was equipped with six eyepieces of different power (100, 140, 180, 200, 310 and 350, producing, respectively, fields of view of 20', 13', 10', 8', 6' and 4'), and that the division of the micrometer screw corresponded to an arc of 0", 314 (Fergola 1865a).

4 Other Instruments Related to Merz

4.1 The Annular Micrometer

The annular (also called circular) micrometer (Fig. 4) by Merz is a filar micrometer equipped with a rack screw which allows for rotation of the field of view. This micrometer is formed by two concentric armillae with internal radii of 17'41", 8 and 8'31", 9, respectively, as per Fergola. The micrometer has a magnification

---

[22] *Macchina di orologeria a pendolo conico. Per regolare la velocità si allunga e si accorcia il pendolino. Ai pesi dati dal costruttore, è convenuto aggiungerne altri per essere altrimenti lo strumento soggetto a fermarsi.*

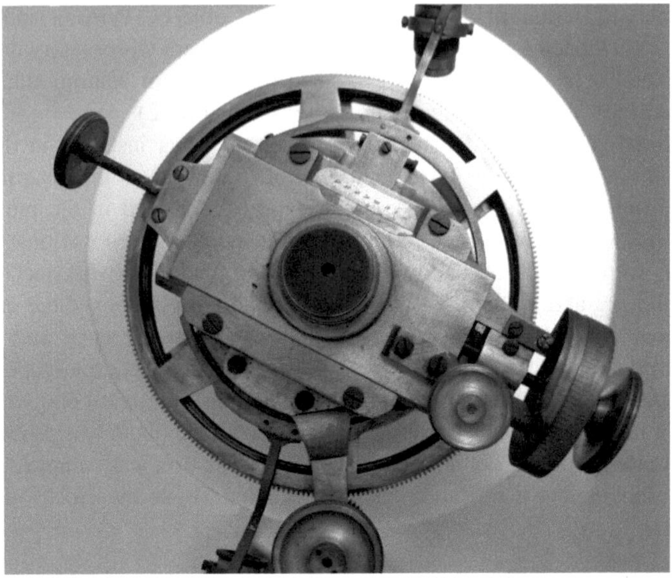

Fig. 4 The annular micrometer by Merz, made in 1863 (MuSA)

power of 50 times and was provided with six eyepieces having magnification power of 40, 60, 90, 140, 200, 310 and 350 times, producing fields of view from 45' to 8'. The graduated scale, bound to the flange, allows for measuring the revolving movement of the armilla by means of two reading verniers. The micrometer screw, leading a graduated cylinder, moves a frame to which three movable wires are secured. A complete rotation of the cylinder, which bears a division into one hundred parts, determines the advancement of one division on the graduated scale placed on the instrument's front face. In order to preserve the set-up when altered by the rotation of the micrometer, there is a screw that allows for an adjustment of the plate perpendicular to the optical axis. The screw is bound to the frame, which houses twelve fixed wires. On the front face of the instrument is located the eyepiece holder, mounted on a slide which can be moved orthogonally to the optical axis by using a rack screw. Besides the observations carried out in February 1865, Fergola used the Merz telescope, equipped with both the filar and annular micrometers, in October, to measure the apparent positions of the newly discovered asteroid Clio[23] (Fergola 1865b).

[23]Clio was discovered by the astronomer Karl Theodor Luther (1822–1900) on the night of August 25th, 1865, at the Bilk Observatory in Düsseldorf.

4.2 The Prismatic Spectroscope

The director of the Florence Observatory, Giovanni Battista Donati, undertook the first stellar spectroscopic studies between 1859 and 1860. He measured the refraction angle and the main absorption lines of fifteen bright stars using a spectroscope designed by himself and manufactured by Poggiali. Some years later, in 1863, the Jesuit Angelo Secchi published an article on his first astronomical spectral observations, carried out by means of a direct-vision spectroscope made by the optician Jean Georges Hofmann (1823–1892) in Paris. In 1865, the director of Capodimonte, De Gasparis, decided to purchase a Donati-Poggiali spectroscope:

> In the coming year, I will definitely purchase the valuable device; indeed, you can consider what I write to you as being an outright commission. Having to be there [in Florence] for some important vote, I shall take advantage of your knowledge and experience to be initiated in the practice of spectroscopic observations.[24] (De Gasparis 1865)

Eight months later, De Gasparis informed Donati of the arrival of the spectroscope from Florence and told him: *I received in good condition the spectroscope [...] I would very much like to see you strongly in Naples in this coming September. You can help us to overcome the initial difficulties regarding the use of the new instrument* (De Gasparis 1866).

The expedition of the Italian astronomers to Sicily to observe the total solar eclipse in 1870, however, persuaded De Gasparis to purchase a new spectroscope, to be used by Nobile, and it was commissioned from the Merz company. Unfortunately, the instrument was not delivered on time, because of the war that had broken out between France and Germany, and Nobile was forced to use the old Donati-Poggiali spectroscope.

Meanwhile, De Gasparis was informed that the Merz spectroscope from Munich was at the Customs of Naples. On December 2nd, the box containing the instrument was delivered to the Capodimonte Observatory and, a few weeks later, De Gasparis received the invoice for *Seventeen Pounds, one Shilling and nine Pence Sterlings*, equivalent to 454.50 Lire, *for the Zölner* [sic!] *direct vision Spectroscope* (*Esercizio* ... 1870). In October 1871, the Pietro Tacchini (1838–1905) astronomer of the Palermo Observatory, was in Naples and tested the new spectroscope by observing some solar prominences; he judged it to be *excellent*.

It is *a direct-vision spectroscope built specifically by Mr. Marz* [sic!] *on the ideas of Zöllner* (Nobile 1872). The spectroscope is a combination of two direct-vision prisms, a slit of adjustable width, and an eyepiece (Fig. 5). Each direct-vision prism consisted of three attached prisms, two made of *crown* glass and a third of *flint* (boron-silicate-lead), which is put between them. This kind of instrument, designed

[24]*Nel venturo anno farò senza fallo acquisto del prezioso apparato, anzi potete ritenere ciò ch'io scrivo come una commissione bella e buona. Toccandomi di essere costì per qualche votazione importante, profitterò de suoi lumi ed esperienza per essere iniziato alle osservazioni collo spettroscopio.*

Fig. 5 The Merz direct vision spectroscope with the slot of three prisms and the original wooden case acquired in 1870, (MuSA)

by the German physicist Johann Karl Friedrich Zöllner (1834–1882), was quite inexpensive and easily fitting onto any type of telescope; it represented a turning point for the development of Italian astrophysics in the mid-nineteenth century, because almost all observatories which were involved in astrophysical research in Italy, at that time, did possess this kind of spectroscope.

4.3 *The Repsold Meridian Circle*

The approach taken by Capocci, respect to the new Italian government, had been successful, providing the Capodimonte Observatory for both additional budget and new instruments. In 1864, Annibale De Gasparis was appointed as director, and continued in the instruments renewal policy. In 1868, he wrote a report to the City of Naples in which he described the conditions of the Observatory's equipment and asked for funds to purchase a new telescope:

> Meridian circle cannot be inverted *[...]* its dark-field wires cannot be illuminated *[...]* the Fraunhofer refractor has been removed, because it was useless on an unstable and oscillating pedestal, and the Observatory now lacks now funds to reassemble it [...] the Reichenbach equatorial telescope *[...]* is now insufficient for researching small planets *[...]* it is not following the daytime movement of the celestial sphere *[...]* Over the world,

> provinces and municipalities in Europe and America compete to encourage and help their scientific establishments. Will our Town Hall remain behind the others, in the case of one of our city's scientific institutes?[25] ("Sulle condizioni ..." [1869])

The Mayor of Naples provided some funds from the City's budget to purchase a new meridian circle, commissioned from the Repsold firm in Hamburg, but the funds covered only about a fourth of the total cost of the instrument (22,596 Lire); the remaining part was paid for with funds allocated by the Minister of Public Instruction, thanks to the support of the Neapolitan historian Pasquale Villari (1827–1917),[26] Secretary of the Minister (Decree 1869). The instrument was delivered in autumn 1871. In December, it was installed in the eastern meridian room of the Observatory on two ash gray granite columns that protected it from any vibrations or movement in the floor (*23 Agosto* ... 1871–1874). The objective-glass of the Repsold meridian circle of the Capodimonte Observatory has a 16.5-cm aperture and a 202-cm focal length and all its optical parts were made by Merz.[27]

The first observation with this instrument was carried out on January 29th, 1874, as recorded in the astronomical registers preserved in the Observatory's Historical Archives. While Schiaparelli was observing Mars with the Brera Merz equatorial telescope (see Carpino, in this volume) and was drafting the large maps of the planet, in Naples, Fergola made many observations of Mars with the Repsold meridian circle;[28] moreover, from observations carried out with the same instrument, Fergola calculated the orbital elements of the great comet of 1881,[29] confirming the good quality of the instrument and its optics (De Gasparis 1881).

---

[25]*Il cerchio meridiano non si può invertire [...] non se ne possono illuminare i fili a campo oscuro [...] Iil Refrattore di Frauenhofer si è dovuto smontare perché inservibile sopra un piede instabile ed oscillante e per rimontarlo mancano i mezzi [...] l'Equatoriale di Reichenbach [...] è oramai insufficiente alla ricerca di piccoli pianeti [...] Questo strumento non segue il movimento diurno della sfera celeste [...] Gareggiano dappertutto Provincie e Municipii in Europa ed in America ad incoraggiare ed aiutare gli stabilimenti scientifici che possiedono, resterà il nostro Municipio indietro agli altri, qualora si tratti d'uno stabilimento scientifico della nostra città?*

[26]Villari took part, together with De Gasparis and Oscarre Capocci (1825–1904), a son of the astronomer, in the 1848 uprisings against the Bourbons and subsequently fled from Naples to Florence. He studied Florentine history and published important political and social essays on the Southern Question. In 1891–1892, he was appointed as Minister of Public Education.

[27]The Adolph Repsold & Söhne company used to order primarily medium and large lenses for telescopes, but also oculars and eyepieces, from the most famous German optical workshops, such as the G. & S. Merz company in Munich. The Repsold and Merz partnership led to a protracted collaboration on some joint projects (Kost 2015).

[28]Fergola observed Mars from August 19th (four days before the first observation of Schiaparelli) to October 23rd, 1877. He also measured the opposition of the red planet in 1879.

[29]The comet was discovered by the Australian astronomer John Tebbutt (1834–1916) on May 22nd. Fergola and his assistants Francesco Contarino (1855–1933) and Filippo Angelitti (1856–1931) observed the comet from June 27th to July 7th, reducing its Right Ascension and Declination from the meridian observations.

Fig. 6 The Repsold meridian circle with Merz objective-lens, in its pavilion, which is covered on the inside with Canada red pine wood

In 1934, the director of the Capodimonte Observatory, Luigi Carnera (1875–1962), decided to build a new pavilion with a steel dome and a motorized opening, realized by the Angelo Bombelli workshop of Milan (Fig. 6). In 1936, the instrument was installed in the pavilion located in the park of the Observatory, to the northwest of the monumental building. For that occasion, the meridian circle was revised in many parts, in particular, the objective was sent to the Carl Zeiss laboratory in Germany for a careful cleaning, as *it had became so opaque that the observations of stars just over the eighth magnitude were very difficult!*. In addition, a mirror was attached at the center of the lens to enlighten the wires in the ocular field by reflection. The eight reading microscopes of the two vertical circles were almost completely reworked by the Officine Galileo in Florence (Carnera 1934).

4.4 The Ertel Theodolite

In May 1815, Reichenbach returned to Munich, after delivering in Naples the new instruments for the Observatory of Capodimonte. Due to some disagreements with Joseph von Utzschneider (1763–1840), Reichenbach left the *Mathematisch-mechanisches Institut* and founded a new workshop with Traugott Leberecht von Ertel (1778–1858), who was engineer of the Institute starting from 1806. In 1821, Georg Friedrich von Reichenbach was appointed as director of the Bavarian Office for Bridges and Roads, and Ertel became the owner of the workshop, changing its name to *T. Ertel & Sohn, Reichenbach'schen mathematisch-mechanisches Institut.* In 1850, Ertel moved the laboratory to Karl Strasse in Munich, close to the workshop of Georg Merz. Thus, the successors of Reichenbach and Fraunhofer continued to stand out in the production of fine mechanics and high quality optical tools. The paths of the two excellent instrument-makers started at Benediktbeuern, in the first decade of the nineteenth century, and then crisscrossed, as they provided components to each other for the construction of the best scientific instruments of the mid-1800s. Therefore, as in the case of the Repsold meridian circle, the Ertel universal theodolite kept at the Capodimonte Observatory has an the objective-glass signed *Merz-Utzschneider und Fraunhofer in München.* The lens was made between 1847 and 1858 (Kost 2015), while the theodolite dates back to the years 1850–1858. It was ordered by the *Commissione italiana per la misura del grado europeo* (Italian Commission for the measurement of the European degree), a board established in 1865 in order to measure the length of the central Europe meridian with accuracy, and therefore to determine, as closest as possible, the geometrical shape of the Earth.[30]

In addition to scientific activities, the Commission decided which instruments would have been provided for each observatory: theodolite, transit instrument, chronometers, chronographs, and barometers (*Commissione* … 1878). The instruments assigned to the Capodimonte Observatory were an universal theodolite by Ertel (Fig. 7), a cylinder cronograph by De Palma,[31] and a telegraph by Mayer & Wolf[32]. The theodolite has a 35-cm azimuth circle, silver graduated and divided into 3', with four verniers which permitted the user to read the azimuth angle within the 2". The graduated limb of the two zenith circles has the same pace, while the

[30]The Commission also included the directors of the Florence, Naples and Brera Observatories, namely Donati, De Gasparis and Schiaparelli. The first meeting was held in Turin in June from the 3rd to the 7th, 1865.

[31]Filippo De Palma (1813-post 1873) was a Neapolitan instrument-maker. He exhibited his electrical machines and physics instruments in Paris in 1867 and in Vienna in 1873.

[32]The telegraph, also known as Oppolzer's instrument, named after astronomer Theodor von Oppolzer (1841–1886), who devised an efficient system for transmitting Morse signals, was commissioned by De Gasparis in February 1875 from the Mayer & Wolf Company in Wien, established in about 1850. The company produced mechanical instruments and electrical devices for electrotherapy, electrolysis and galvan-acoustics until about 1945. See www.beniculturali.inaf.it/sicap/opac.aspx?WEB=INAFS&TBL=PST&ID=399.

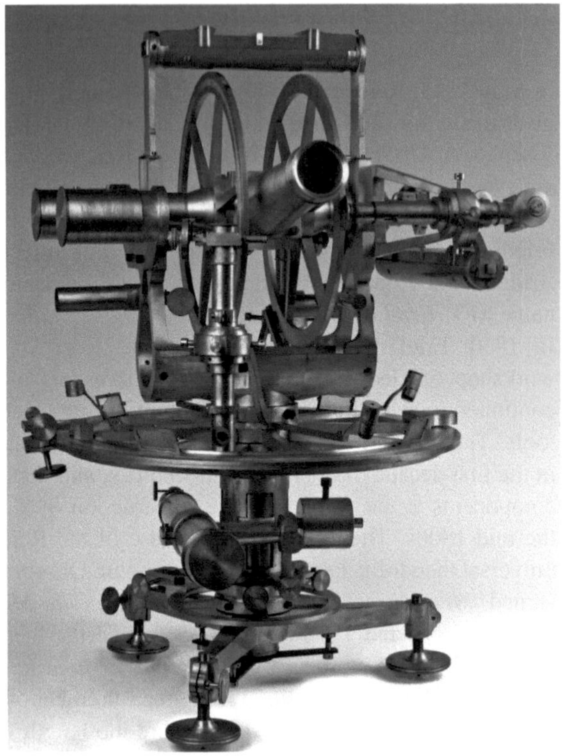

Fig. 7 The Ertel universal theodolite with a cropped tail telescope invented by Reichenbach, called *Stutzschwanz* (MuSA)

vernier of each circle is within 1". From 1903 to 1909, the instrument was lent to the geodesy chair of Messina University. Because of the 1908 earthquake, it suffered some damage and, two years later, the University of Messina took responsibility for the costs to repair *the Ertel alt-azimuth telescope by the talented mechanic of this Observatory, Mr. Pasquale Moreno, on behalf of the Italian Geodetic Commission* (University ... 1910).

4.5 The Small Merz Telescope

The last instrument of the collection is another Merz telescope, called by the astronomers of Capodimonte "the small Merz". The telescope has a mahogany tube and brass housings for the objective and the ocular; both lenses are missing. It is signed *G. & S. Merz in München* (Fig. 8). It seems to be an instrument dating back to the second half of the nineteenth century; however, neither historical documents held in the Observatory archives, nor any papers, essays, or communications between astronomers, ever mention the" small Merz".

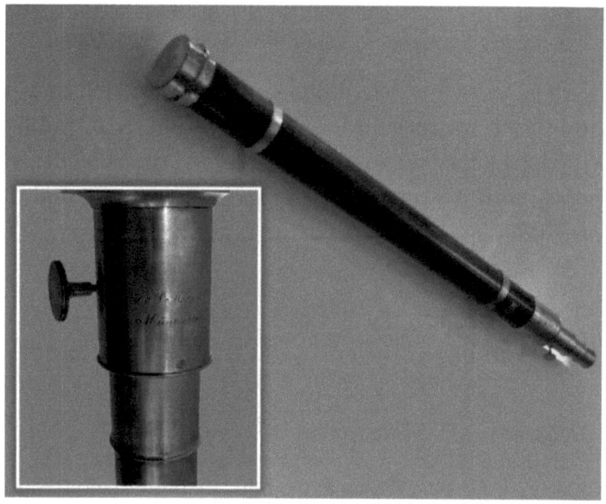

Fig. 8 The "small Merz," telescope with a detail of the signature engraved on the ocular housing (*bottom-left*) (MuSA)

In a letter from 1862, the Merz Company reminds Capocci about a missing payment of 273 florins for *the two telescopes provided to the Royal Observatory of Naples in March 1847* (Merz 1862b). Is the small Merz one of these two telescopes? Why is neither telescope ever mentioned in inventories and reports by Capocci and his successors?[33]

The first information on this instrument has been found in a modern inventory register. It is identified as a *Merz Telescope, with a wooden tube, a rack, seven eyepieces, and an equatorial mounting*; below that, there is a note that was penciled in later: *in the workshop in two pieces* (*Registro* … 1948–1979). The objective was 11 cm in aperture with a focal distance of 160 cm, but presently, the optical parts are missing. The only scientific work mentioning this telescope is dated 1959 and regards to the observations of a solar eclipse. The astronomer Tito Nicolini (1899–1981) took some pictures of the phenomenon:

> October 2, [1959] on the occasion of the solar eclipse, that was partial in Italy, I have observed it with the little equatorial Merz telescope (11 × 160 cm), realizing 49 photographs on 35 mm film. Dr. Vassallo made the critical measure with the Askania comparator of the Observatory, and the Abbe comparator of the Physics Institute (Prof. Carrelli [(1900–1980), director of the Physics Institute] made it available, kindly), and he determined the contact times. Dr. Fichera also took part in the photographic observations of the eclipse using the 13 × 200 Merz telescope, and Dr. Vassallo made visual observations with the Dollond telescope equipped with the Lyot filter.

[33]One month after the date of this letter, Capocci wrote to the Minister of Public Instruction to ask permission to pay the remaining amount for the new Merz telescope, but he did not mention the request made by Merz.

> In summary, we found that the most accurate instrument for measuring the contact times is definitely the latter, as would be expected.[34] (Nicolini 1960)

One year later, Nicolini again used the "small Merz" to take thirteen photographs during the transit of Mercury across the Sun, publishing a photo in the *Annuario* (1961). In the subsequent reports by the director, the "small Merz" again disappears, until it was recently recovered, restored and displayed in the new exhibition of MuSA.

References

"23 Agosto 1871–17 Settembre 1874 Repsold", 1871–1874, Observation logbook, HAOC, ASO, B. 2, f. 6.

Bemporad, A., 1919, Letter to the Ministry of Education, Naples, December 1, HAOC, ASS, B. 1, f. 8.

Bianchi, S., 2012, "Un Imperatore ad Arcetri", *Giornale di Astronomia*, 38(2), 2–12.

Brenni, P., 1994, "19th century French scientific instrument makers, III: Lerebours et Secretan", *Bulletin of the Scientific Instrument Society*, 40, pp. 3–6.

Brioschi, C., 1824–26, *Comentarj astronomici della Specola Reale di Napoli*, Napoli, Tipografia nella Pietà de' Turchini.

Capaccioli, M. and Gualano, S., 2012, *Arminio Nobile e la misura del cielo*, Milano, Springer.

Capocci, E., 1861a, Letter to Francesco De Sanctis, Turin, March 30, SAN, MPI, B. 753/II, f. 19.

Capocci, E., 1861b, Draft of a letter to Francesco Brioschi, Naples, December 3, HAOC, SLF, B. 3, f. 2.

Capocci, E., 1861c, Letter to Francesco De Sanctis, Naples, December 3, SCA, MPI.

Capocci, E., 1862a, Draft of a report: ("materiale"), HAOC, SLF, B. 3, f. 2.

Capocci, E., 1862b, Draft of a letter to Carlo Matteucci, Naples, [September 10], HAOC, SLF, B. 3, f. 2.

Carnera, L., 1934, "La Specola di Capodimonte negli anni 1932–33", *Calendario del R. Osservatorio Astronomico di Napoli*, 58.

Commissione italiana per la misura del grado europeo, 1878, *Processo verbale delle sedute della Commissione italiana per la misura dei gradi, prima riunione dell'anno 1865 in Torino dal 3 al 7 giugno*, Firenze, Tipografia di G. Barbera.

Decree of the Minister of Public Instruction, 1869, Florence, August 17, HAOC, ASS, B. 1, f. 13.

de Gasparis, A., 1861, Letter to Quintino Sella, Naples, May 20, Historical Archives of Fondazione Sella, Carteggio.

de Gasparis, A., 1865, Letter to Giovanni Battista Donati, Naples, November 28, HAOA, FA, GBD, Correspondence.

de Gasparis, A., 1866, Letter to Giovanni Battista Donati, Naples, July 18, HAOA, FA, GBD, Correspondence.

---

[34]*Ed il 2 ottobre al piccolo equatoriale Merz (11 × 160 cm) [ho] eseguita, in occasione della eclisse di Sole osservata come parziale in Italia, una serie di 49 fotografie, su pellicola di 35 mm. La delicata misura al Comparatore Askania dell'Osservatorio, ed al comparatore Abbe dell'Istituto di Fisica (gentilmente posto a disposizione dal Prof. Carrelli), e la riduzione per la determinazione degli istanti di contatto, sono state affidate al Dott. Vassallo. Alla osservazione fotografica della eclisse, prese parte anche il dott. Fichera, al Merz 13 × 200 cm, e lo stesso dott. Vassallo, visualmente, al Dollond munito di filtro di Lyot. All'atto pratico si è riscontrato, come prevedibile, che la determinazione pi precisa dell'istante di contatto è nettamente quest'ultima.*

de Gasparis, A., 1881, "Eléments et observations méridiennes de la Grande Comète (b) 1881", *Astronomische Nachrichten*, 100, 125–126.
de Gasparis, A., 1884, *Sullo stato del R. Osservatorio di Capodimonte e sui lavori in esso eseguiti dal 1819 al 1883*, Napoli, Tipografia dell'Accademia Reale delle Scienze, 8.
De Sanctis, F., 1861, letter to Giovanni De Sanctis, [Turin], February 19, Talamo, G. (ed.) *Epistolario (1861–1862)*, Einaudi, Torino, 1969; p. 6.
Del Giudice, F., 1865, "De' lavori accademici dell'anno 1864", *Atti del R. Istituto d'Incoraggiamento alle Scienze Naturali Economiche e Tecnologiche di Napoli*, S. 2, 2, 25.
"Esercizio 1870", 1871, Account book for the year 1870, January 7, HAOC, Amministrazione, Conti, B. 5, f. 2.
Fergola, E., 1865*a*, Determinazione degli errori costanti dell'equatoriale di Merz esistente nella R. Specola di Napoli, Rendiconto dell'Accademia delle Scienze Fisiche e Matematiche, 4(1), 119–124.
Fergola, E., 1865*b*, "Osservazioni ed elementi dell'orbita del pianeta Clio", *Rendiconto dell'Accademia delle Scienze Fisiche e Matematiche*, 4(10), 315–316.
Gargano, M., 2016, "Carlo Brioschi, il primo astronomo di Capodimonte", *Giornale di Astronomia*, 42(2), 10–16.
Gargano, M., 2013, "Arminio Nobile", *Polvere di Stelle.*
Gargano, M, E. Olostro Cirella and M. Della Valle (eds.), 2015, *Che il diavolo benedica i Pulcinella!: cronache napoletane, scientifiche e non, dell'astronomo von Zach*, Napoli, Tullio Pironti editore.
Kost, J., 2015,*Wissenschaftlicher Instrumentenbau der Firma Merz in München (1838–1932)*, Hamburg, Tredition.
Lacchini, G.B., 1919, Letter to Azeglio Bemporad, Faenza, December 29, HAOC, Corrispondenza Privata, Azeglio Bemporad, B. 1, f. 1.
Lombardi, A., Mandrino, A., 2008, "Ricerca, istruzione e divulgazione all'Osservatorio astronomico di Brera" in Canadelli, E. (ed.) *Milano scientifica 1785–1924*, vol. 1: La rete del grande Politecnico, Milano, Sironi, 95–116.
Merz, G. & S., 1862a, Letter to Ernesto Capocci, Munich, July 10, HAOC, ASS, B. 1, f. 8.
Merz, G. & S., 1862b, Letter to Ernesto Capocci, Munich, September 5, HAOC, ASS, B. 1, f. 8.
Nicolini, T., 1960, "Relazione sull'attività dell'Osservatorio di Capodimonte nel 1959", *Annuario dell'Osservatorio Astronomico di Capodimonte* - Napoli, 71–75.
Nobile, A., 1875, *Saggio di un nuovo metodo per l'osservazione delle distanze scambievoli delle stelle multiple*, Napoli, Tipografia della R. Accademia delle Scienze Fisiche e Matematiche.
Nobile, A., 1872, "Osservazioni sull'eclisse totale di Sole del 22 dicembre 1870 fatte in Terranova di Sicilia" in Cacciatore, G. (ed.) *Rapporti sulle osservazioni dell'ecclisse totale di Sole del 22 dicembre 1870 eseguite in Sicilia dalla Commissione italiana,* Palermo, Stabilimento Tipografico Lao, 121–123.
Olostro Cirella, E. & Gargano, M., 2016, "The solar eclipse of 1870" in Chinnici, I. (ed.) *Starlight. La nascita dell'astrofisica in Italia*, Napoli, Arte'm, 56–65.
Perrotta, F., Gargano, M., 2016, "I viaggi dell'equatoriale di Merz: le vicende umane e scientifiche che hanno accompagnato il telescopio a Napoli, a Terranova e a Faenza", *Atti del XXXIII Convegno SISFA*, held in Acireale and Catania 4–7 September 2013, (in press).
Registro cronologico delle operazioni inventariali, 1948–1979, HAOC, Amministrazione, Inventari e atti di Consegna, Inventari, 2.
Sella, Q., 1861, Letter to Paolo Emilio Imbriani, Turin, March 17, SAN, MPI, B. 753/II, f. 19.
"Stato discusso del dicastero dell'Istruzione pubblica nelle Province Napoletane, per l'anno 1861", 1861, February 10, SAN, MPI, B. 753/II, f. 2.
"Sulle condizioni della R. Specola di Capodimonte", [1869], report of astronomers to the Naples Administration, HAOC, Amministrazione, Relazioni, B. 1, f. 1.
University of Messina, 1910, Restitution of the universal theodolite, HAOC, ASS, Varie, B. 3, f. 5.
Visconti, F., 1832, "Breve compendio delle opere astronomiche presentate alla Reale Accademia delle Scienze da' Socj Signor Brioschi, Direttore della Specola Reale di Napoli, e Signor Cacciatore Direttore della Specola Reale di Palermo; cio del volume I. de' Comentarj della

Specola Reale di Napoli, e del volume I. del Reale Osservatorio di Palermo", *Atti della Reale Accademia delle Scienze, sezione della Società Reale Borbonica*, III, 37–56.
Visita del Ministro della pubblica Istruzione, 1862, *Giornale storico degli Archivi Toscani* 6(2), 152.
Zach, F. von, 1819, ***Correspondance astronomique**, géographique, hydrographique et statistique*, vol. 2, p. 537.

A Merz Telescope on Mount Etna: The Catania Astrophysical Observatory

Andrea Orlando

1 Introduction

After the establishment of the Palermo Astronomical Observatory, at the end of the 18th century, some attempts were also made to establish an astronomical observatory in Catania. The Etna's city was home to one of the oldest universities in Italy, but only theoretical astronomy was taught there, privileging the mathematical approach and neglecting the practice, on account of the University's lack of appropriate astronomical instruments. More for political than for scientific reasons, these attempts failed, and for many years, the idea of having an astronomical observatory in Catania was abandoned. In the last quarter of the 19th century, however, thanks to the impetus of Pietro Tacchini (1838–1905), astronomer at the Palermo Observatory and a pioneer in solar physics research, the project of building an astronomical observatory was relaunched: Tacchini was interested in the study of solar corona and therefore intended to build the Observatory on the high slopes of Mount Etna, in order to obtain the best possible visibility, minimizing the effect of the earth's atmosphere (Tacchini, 1876).

A. Orlando (✉)
INAF-Osservatorio Astrofisico di Catania, Catania, Italy
e-mail: aorlando@oact.inaf.it

A. Orlando
National South Laboratories (INFN), Catania, Italy

A. Orlando
Institute of Sicilian Archaeoastronomy, Novara di Sicilia, Italy

I. Chinnici (ed.), *Merz Telescopes*, Historical & Cultural Astronomy,
DOI 10.1007/978-3-319-41486-7_8

1.1 The Astronomical Stations in the Etna Area

In 1876, a solemn public meeting of the Gioenia Academy, an ancient and prestigious scientific institution in Catania, was held on the occasion of the ceremonial carrying of the ashes of the famous musician Vincenzo Bellini (1801–1835) to Catania. During the meeting, Tacchini revived the proposal for the construction of an astronomical observatory, to be built on Mount Etna.[1]

The new observatory was intended to include the "British House" (*Casina degli Inglesi*), the refuge embodying the "Gratissima" shelter, erected by the geologist Carlo Gemmellaro (1787–1866) in 1804 (Fig. 1). In honor of the great composer from Catania, the building would have been called the *Bellini Observatory* (Millosevich et al. 1880; Riccò et al. 1898). The proposal was welcomed by the government, which, by the Royal Decree of February 17th, 1878, entrusted the overall direction of the construction work to Tacchini.[2]

The astronomer then emphasized the necessity of having another observatory in the city to continue the astronomical observations during the winter season, when the astronomical station on Etna became almost inaccessible. The City of Catania accepted the proposal and allotted several rooms of the ex-Monastery of the Benedictines for the additional observatory. In 1885, the construction of the astronomical station at Catania was decreed. In 1886, during a presentation of essays regarding celestial photography at the Academy of the Lincei, Tacchini noted that the observatory in Catania, because of its climate and location, was perfectly suited to take part in the *Carte du Ciel* project, the international endeavor proposed by the French Academy of Sciences to photograph the entire starry sky.

This proposal somehow modified the initial scientific program, as the Catania and the Etna Observatories were mainly destined to contribute to physical astronomy rather than astrometric studies. For this reason, the first Italian astrophysics tenure was established at the University of Catania in 1890 and assigned to Annibale Riccò (1844–1919), former astronomer at the Palermo Astronomical Observatory (Abetti 1920).

[1]The assembly took place on September 22nd, 1876; in his speech, entitled '*Della convenienza ed utilità di erigere sull'Etna una stazione astronomica-meteorologica*', Tacchini described a promising observational experience that he had on Etna on September 18th, 1876. Indeed, a letter sent from Tacchini to the Mayor of Catania for proposing to build an astronomical station on Mount Etna, is dated September 19th, 1876 [the letter is preserved in the Historical Archive of the Catania Astrophysical Observatory in *Corrispondenza,* b.1, fasc. 2 (AMI)]. A few years earlier, in 1871, Tacchini had already, unsuccessfully, proposed to the Ministry of Education the setting up of an observatory on the volcano Etna.

[2]The Royal Decree of 4352 (Series 2) of February 17th, 1878 is kept in the Catania State Archives.

Fig. 1 The *Casina degli Inglesi* and the *Gratissima* (Wolfgang Sartorius von Waltershausen 1835)

1.2 Bellini Observatory. A Merz Telescope on Mount Etna

The restyling of the *Casina degli Inglesi* was completed in the second half of 1879. In September of the same year, the iron 8-m diameter dome, built by the Fonderia Oretea in Palermo, was also ready to be assembled. However, due to bad autumn weather, the installment didn't take place until 1880 (Fig. 2).

The dome of the Bellini Observatory was meant to house a refractor telescope produced by the Merz Company of Munich in Bavaria, which was a leader in the field of optics. Tacchini had already sent a request for the telescope in 1876;[3] in response to that letter, the Merz Company generously offered a 12-inch lens, at the price of a 10.5-inch lens. The cost of the refractor telescope with a lens of 34-cm aperture was approximately 12,500 liras.[4] In a document dated July 16th, 1878, the

[3]In a letter dated October 20th, 1876, written by Tacchini to Giuseppe Lorenzoni (see Zanini, in this volume), the former expressed his idea of buying two tubes from Merz (HAOPd, Lorenzoni Fund, correspondence).

[4]From the *Rendiconto delle anticipazioni ricevute e delle spese sostenute per l'Osservatorio Bellini sull'Etna a tutto il 17 del mese di giugno* (State Archive of Catania), it is possible to presume that the amount was divided into two payments: a 5401 liras deposit (paid on August 17th, 1877) and a balance of 7140 lira (paid on April 24th, 1878).

Fig. 2 The Bellini Observatory on Mount Etna, in a picture dated 1892 (INAF-OACt Archive)

G. & S. Merz Company of Munich acknowledged the receipt of the first payment (Fig. 3).

Tacchini purchased the refractor (557-cm focal length) from Merz, but commissioned its parallactic mounting (Fig. 4) from the expert technician Giuseppe Cavignato,[5] who worked at the Padua Astronomical Observatory. An iron tube was installed at Bellini Observatory, but, unfortunately, it has been lost, and there is no record of the manufacturer (it is possible that it was the Fonderia Oretea, which built the dome, but no evidence of this has yet been found). The Merz visual doublet (34-cm diameter) was formed by a steel barrel with a bronze front collar on which were engraved the words: '*G. & S. Merz in Muenchen*'. The objective-lens corrected the chromatic aberration in the spectrum region between the Fraunhofer lines F (4861 Å) and C (6563 Å). The Merz lens is now kept at the headquarters of the Catania Astrophysical Observatory (Fig. 5), where some historical items are on display.

In an article, Annibale Riccò described the details of the refractor telescope:

> … right ascension circle, which is read with two small telescopes, giving the 10" directly and estimating the ½ 1"; other right ascension circle, which is read from the lower end of the polar axis in pointing the instrument and gives with the vernier 5"; declination circle that reads from the end eyepiece with two telescopes and gives the 2'. Clockwork movement to follow the stars, with a wings regulator with a variable skew according to speed. Seeker with objective of 0.06 m of diameter, focal length of 0.62 m. Wire micrometer built

[5]With a contract signed on February 3rd, 1879, the mechanic Cavignato agreed to build two parallactic mountings for a total cost of 10,750 liras. The first mounting was completed in the second half of 1881 and the other in the following year.

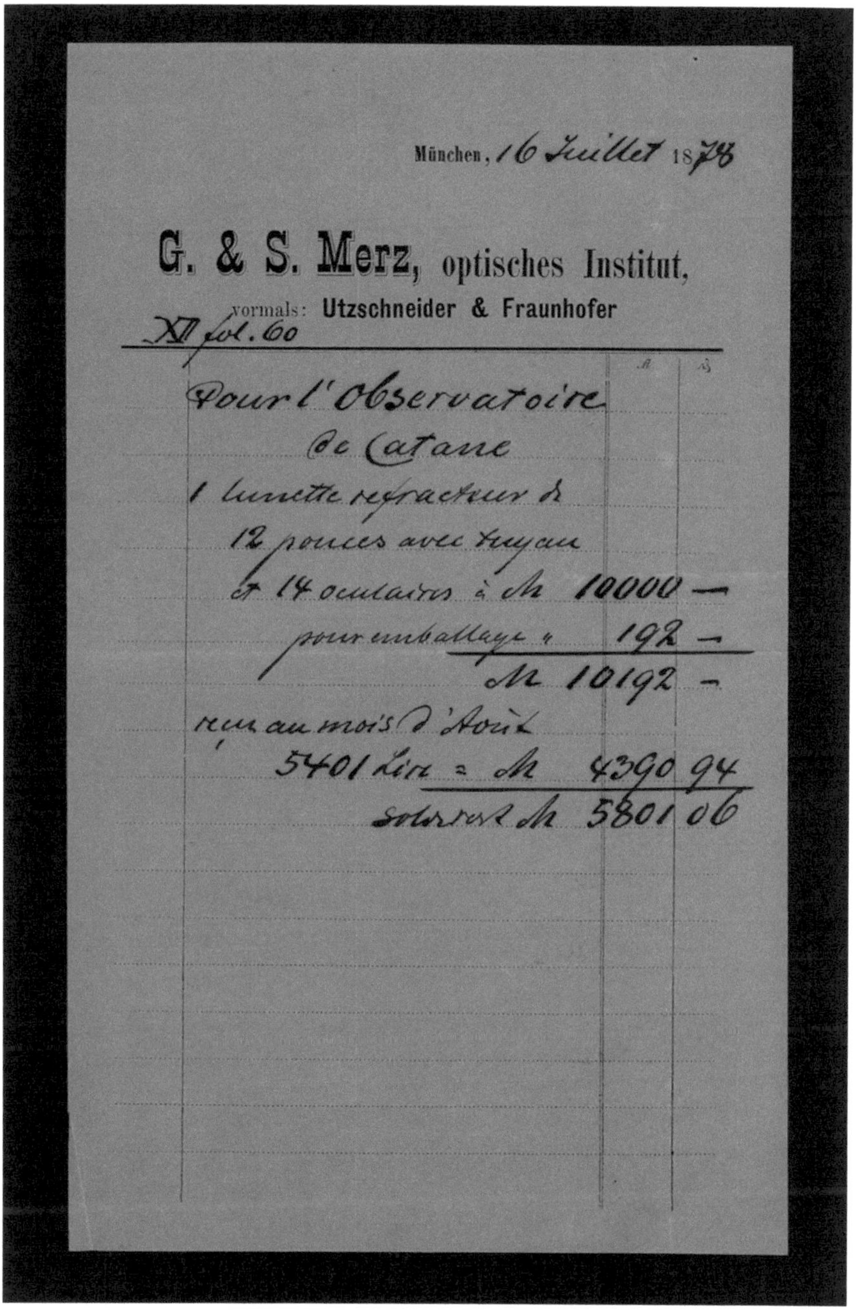

München, 16 Juillet 1878

G. & S. Merz, optisches Institut,
vormals: Utzschneider & Fraunhofer

XII fol. 60

Pour l'Observatoire
de Catane
1 lunette refracteur de
12 pouces avec tuyau
et 14 oculaires à M 10000 —
pour emballage " 192 —
M 10192 —
reçu au mois d'Août
5401 Lire = M 4390 94
solde rest M 5801 06

Fig. 3 Document regarding the Catania refractor by Merz (courtesy of State Archive of Catania)

Fig. 4 The Merz refractor telescope in two archival images showing the instrument, with its mounting, during testing in the *Hall of Figures* (*Sala delle Figure*) at the Padua Astronomical Observatory (INAF-OACt Archive)

Fig. 5 The 34-cm diameter Merz lens in the scientific collection of the Catania Astrophysical Observatory (INAF-OACt collection)

by Salmoiraghi: with micrometer screws in the right ascension and declination, and a position circle that gives the 1 ½′… the tube that is made of iron (while the one at Catania Observatory is made of wood) bears two sleeves flowing along two guides, upon which another telescope can easily be mounted, or even a simple objective with the aid of a sleeve of black canvas which serves as a tube. With these adaptations, one or the other of the two optical lenses that the Observatory possesses can be readily mounted on the instrument.[6] (Riccò et al. 1898)

6 *... cerchio di ascensione retta, che si legge con 2 cannocchiali dall'estremità oculare e dà direttamente i 10", ed a stima il ½ 1"; altro cerchio di ascensione retta, che si legge dall'estremità inferiore dell'asse polare nel puntare lo strumento e dà col nonio i 5"; cerchio di declinazione che si legge dall'estremità oculare con due cannocchiali e dà i 2'. Movimento di orologeria per seguire gli astri, con regolatore ad ali di obliquità variabile secondo la velocità. Cercatore coll'obiettivo del diametro di metri 0.06, lunghezza focale di metri 0,62. Micrometro filare costruito da Salmoiraghi: con viti micrometriche in ascensione retta a declinazione, e cerchio di posizione che dà il ½ 1'.... il tubo che è di ferro (mentre quello dell'Osservatorio di Catania è di legno) porta due manicotti scorrenti lungo due guide, nei quali si può facilmente montare un altro cannocchiale, od anche un semplice obiettivo col sussidio di una manica di tela nera che fa l'ufficio di tubo. Con questi dispositivi si può montare prontamente sullo strumento l'uno o l'altro dei due obiettivi ottici che l'Osservatorio possiede.*

From the same article, it is also possible to deduce some important information about the installation of the instrument:

> On August 17th, 1891, the mechanic goes to the observatory to install the other equatorial mounting for the 0.34 m lens; the cleaning of the pieces and the placement lasts until September 27th of the same year. At the same time, at the Etna Observatory, substantial repairs are executed to address the faults that occurred as a result of the bad weather and the violent earthquakes of 1886.[7]

In addition, Riccò recalls that:

> On October 25th, 1892 (the fury of the great eruption of Mount Etna, which lasted from July 8th to December 29th, having somewhat diminished), the lens is brought to the Etna Observatory and applied to the equatorial mounting, the inclination and azimuth adjustments are made, the instrument is tested on various celestial objects, and one remains satisfied with the clear and beautiful images that it is possible to observe.[8]

There is no extant image of the refracting telescope installed at the Bellini Observatory, and the known photographs of the observatory on Mount Etna show the exterior of the building.[9] However, in June 2015, during the work to digitize some photographic plates, dating from the late 19th and early 20th centuries and kept at the Catania Astrophysical Observatory, an extremely rare photograph was found: it shows one of the rooms on the upper floor of the Bellini Observatory,[10] and a Huggins photographic coronagraph can be seen in the foreground (Fig. 6).

As it turned out, the unfavorable weather conditions and volcanic activity on Mount Etna would never permit the astronomical observations at Bellini Observatory to be carried out regularly, so that, as a result, all scientific activities were soon being carried out entirely at Catania. In fact, in 1925, the Etna Observatory was given to the Institute of Volcanology, which used it until its complete destruction by a lava flow that occurred during an eruption in 1971. It is assumed that, prior to or after the date of the transfer of the Bellini Observatory to

[7] *Al 17 agosto 1891 il meccanico si reca all'osservatorio per collocare l'altra montatura equatoriale per l'obiettivo di metri 0.34; la ripulitura dei pezzi e la collocazione dura fino al 27 settembre dello stesso anno. Nello stesso periodo di tempo si eseguiscono all'Osservatorio Etneo notevoli riparazioni ai guasti che aveva sofferto per le intemperie e per i violenti terremoti del 1886.*

[8] *Il 25 ottobre 1892 (essendo alquanto diminuito il furore della grande eruzione dell'Etna, la quale durò dall'8 luglio al 29 dicembre), si porta all'Osservatorio dell'Etna l'obiettivo, lo si applica a quella montatura equatoriale, si fa la rettifica dell'inclinazione e dell'azimut della medesima, si prova lo strumento sopra varii oggetti celesti, e si resta soddisfatti delle immagini tranquille e splendide che si ottengono.*

[9] One of the richest collections of photographs regarding the Bellini Observatory is certainly the one owned by the volcanologist Gaetano Ponte (1876–1955), now deposited in a fund at the Tuscan Photographic Archive.

[10] A detailed description of the Bellini Observatory can be found in two documents: a report on the inspection of the premises and the equipment of the Observatory prepared by the Royal Corps of Civil Engineers on September 26th, 1886, and a delivery record for the keeper prepared by the Prefecture of Catania Province on October 3rd, 1889. Both documents are stored at the State Archives of Catania.

Fig. 6 The Huggins photographic coronograph in the only extant photograph portraying part of the great round room on the upper floor of the Bellini Observatory (INAF-OACt Archive)

the volcanologists, the Merz telescope and its mounting, and, of course, all the other instruments stored at the mountain astronomical station, were moved to Catania. However, today, any trace of this important instrument related to the establishment of the Catania Astrophysical Observatory has been lost.

1.3 The Merz Telescope at the Catania Observatory

As early as 1880,[11] it was agreed that a room would be set up in Catania to install one of the two parallactic mountings ordered by Pietro Tacchini from Cavignato. For the construction of the additional observatory in the town, the circular room, which was the refectory of the ex-Monastery of the Benedictines, was chosen. Above this room, another circular room was built, with a rotating 8-m diameter dome, made by the Audisio Company from Turin.

From Riccò's writings (Riccò et al. 1898, p. 146), we learn that the transportation of the refractor to the new circular room, the assembly of the instrument and the preparation of all related accessories started in early January 1891 (Figs. 7 and 8):

> On January 1st, 1892, the daily observations of the spots, faculae, and solar prominences began, and were then regularly published in the Memoirs of the Italian Spectroscopists' Society and the Astronomische Mittheilungen of Zurich for what concerns the statistical of the sunspots only.[12]

With the Merz telescope in Catania, as well as with its twin on Mount Etna, solar spectroscopic observations were conducted, as originally planned by Tacchini, and, in order to improve the quality of this kind of observation, a spectroheliograph was acquired. This magnificent instrument, designed by Riccò and George Ellery Hale (1868–1938), was built by *Officine Toepfer & Sohn* in Potsdam and delivered to the Catania Observatory at the end of 1906 (Riccò et al. 1908). The spectroheliograph, weighing about 45 kg, was attached to the Merz tube with special adjustments, in order not to overload the barrel of the telescope too much. From early June 1908, regular and daily photographs of the photosphere and chromosphere were taken; this activity lasted for over thirty years.

The Merz telescope installed in the downtown dome was used for at least sixty years, before major maintenance was carried out (Fig. 9). From the Catania Astrophysical Observatory yearbooks, it is evident that the last major maintenance campaign dedicated to the 'surviving' Merz telescope was conducted in the years 1946–49 under the directorship of Eugenio De Caro (1899–1954).[13]

---

[11]On July 30th, 1880, an agreement was signed between the Ministry of Agriculture, Industry and Commerce and the Ministry of Education for the establishment of the new Observatory. On July 19th, 1882, a resolution was made by the City Council of Catania to allot some of the premises of the former Monastery of the Benedictines to host the downtown branch of the Bellini Observatory. The final agreement, signed on December 31th, 1883, by the Ministries of Education and Agriculture, Industry and Commerce, the Province, the City and other related institutions, for the establishment of an astronomical and meteorological observatory in the ex-Monastery of the Benedictines was ratified by the Royal Decree 2803 dated October 5th, 1884.

[12]*Al 1° gennaio 1892 si sono cominciate le osservazioni quotidiane delle macchie, facole, protuberanze solari, che poi si sono pubblicate regolarmente nelle Memorie della Società degli Spettroscopisti Italiani e nelle Astronomische Mittheilungen di Zurigo per ciò che riguarda solo la statistica delle macchie solari.*

[13]Catania Astrophysical Observatory Yearbooks, 1948–53.

Fig. 7 The Merz refractor telescope at the Catania Observatory with the Hale-Riccò spectroheliograph attached in the focal plan (INAF-OACt Archive)

Fig. 8 Details of the furnishings of the room hosting the Merz telescope (INAF-OACt Archive)

Fig. 9 The Merz-Cavignato equatorial, dismantled during the 1946–49 maintenance work (INAF-OACt Archive)

The original site of the Catania Astrophysical Observatory, in the historic center of Catania, was abandoned around 1965, when the headquarters were moved to the current site.[14] As a consequence of the move, the old telescopes were left in a condition of increasing neglect and degradation. The instrument once located on Mount Etna is today completely untraceable, while the Merz telescope at the town station had a slightly different story. In the early 1970s, under the directorship of Giovanni Godoli (1927–2006), the telescope and its mounting were lent to an association of amateur astronomers in the Marche region. About forty years later, during the directorship of Giovanni Strazzulla, the wooden tube of the instrument was recovered, and today is kept in one of the warehouses of the Catania Astrophysical Observatory. It is hoped that the Cavignato mounting may also soon be recovered and that the remaining parts of the Merz telescope can be restored and properly placed in a museum.[15]

---

[14]The Catania Astrophysical Observatory is today located in the University Campus.

[15]On July 9th, 2015, in the west classroom of the Catania Astrophysical Observatory, the author of this chapter held a seminar on the Catania Astrophysical Observatory's historical instruments. During the seminar, he suggested the recovery of some borrowed instruments and the restoration

2 Other Telescopes of the Late 19th and Early 20th Centuries

The town station of the Royal Catania Astrophysical Observatory, in addition to possessing the observatory above the anti-refectory of the ex-Monastery of the Benedictines, and occupying some rooms in the northern part of the building, was equipped with a series of pavilions in the garden (Fig. 10) that housed a number of astronomical instruments of great scientific value for the time. In the 1960s–1970s, the garden, with all the pavilions, was sold to the Vittorio Emanuele Hospital, which then demolished the structures to build new hospital rooms. Below, the most important astronomical instruments that were housed in the halls of the Catania Astrophysical Observatory are briefly mentioned and described.

Under Tacchini's urging, in 1889, the Catania Observatory was included on the list of the observatories participating in the *Carte du Ciel*. Consequently, it was equipped with an equatorial photographic telescope (Fig. 11). The lens of this 33-cm aperture instrument (focal length of 3.46 m) was built by Steinheil in Munich, while the equatorial mounting with a 21-cm visual telescope (focal length of 3.18 m) was made by Salmoiraghi of Milan; a wire micrometer, two bronze chassis and an apparatus for direct magnification were constructed by Gautier in Paris (Riccò et al. 1898, p. 154).

The telescope was installed in 1891 in a pavilion, built in the Benedictine Monastery Garden; the Royal Observatory of Catania was becoming larger, covering an area of approximately 4000 m^2. The rotating dome of the pavilion hosting the telescope, dedicated to the *Carte du Ciel* project, had a diameter of 5.5 m and was built by the Audisio Company.

The grueling work of observing and photographing the sky occupied the Catania Observatory for more than fifty years. As a matter of fact, the Carte du Ciel catalogue was completed in 1942, in the midst of World War II, and it is important to emphasize that the Catania and Helsinki Observatories were the only ones to complete their assigned parts of the catalogue work as originally planned (Chinnici et al. 1999), most especially thanks to the effort of Azeglio Bemporad (1875–1945) (Mangano et al. 2013).

An equatorial telescope built by Cooke in New York, presumably between 1880 and 1882, was commissioned by Pietro Tacchini to observe the total solar eclipse of May 6th, 1883, in Micronesia (Tacchini 1888). After that expedition, the instrument was sent to the Catania Observatory, where it was placed in a small circular pavilion, 4.75-m in diameter, covered by a wooden cylindrical dome (Fig. 12). The

(Footnote 15 continued)

of the ancient telescopes, including the Merz, in order to create the 'Museo della Specola di Catania (MUSC)' on some available premises of the ex-Monastery of the Benedictines, the place where the town station of the Catania Observatory was originally established at the end of the 19th century.

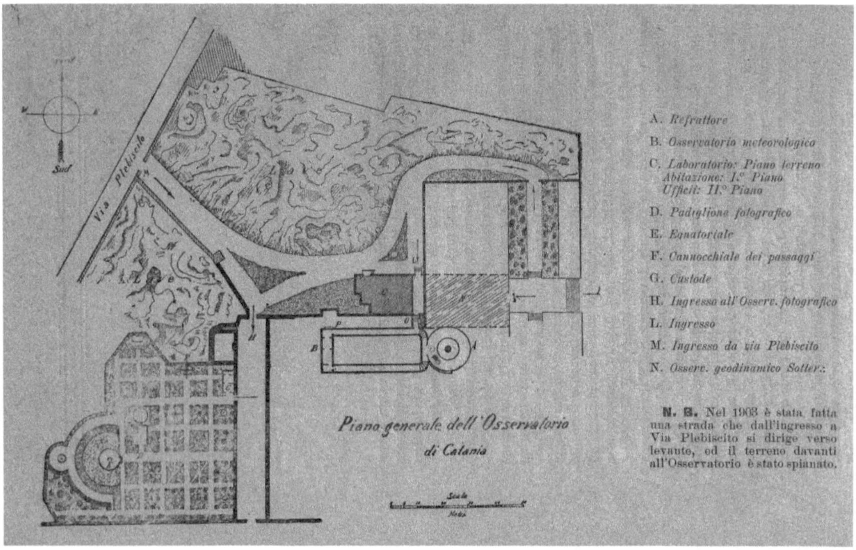

Fig. 10 Plan of the downtown site of the Catania Astrophysical Observatory (from: *Gli Osservatori di Catania e dell'Etna*, La Sicilia Universitaria Editrice, 1905)

Cooke equatorial was mostly devoted to the observation and statistics of solar activity phenomena (spots and faculae), and to the spectroscopic observation of the solar limb (bumps and height of the solar chromosphere). In the early 1970s, the instrument was given on loan to the Astronomers Club of Catania. It is to be hoped that this important instrument might be recovered as soon as possible, so as to be restored and properly displayed in a museum.

In early July 1892 started the renovation of the pavillon for housing the transit instrument loaned from the Collegio Romano Observatory. The instrument (having a 78-mm diameter objective and 117-mm focal length, equipped with two positive eyepieces, one of them for zenithal observations) was built by Reichenbach in Munich and served mainly for time determination. In 1914, the transit instrument was replaced by the Ertel meridian circle (focal length = 1.5 m, aperture = 9 cm) (Fig. 13), once used by Angelo Secchi (1818–1878) at the Collegio Romano Observatory. Today, the instrument is at the Cagliari Astronomical Observatory, given by Godoli on permanent loan in 1971.

Many other interesting instruments, listed below, have been used at the Catania Astrophysical Observatory:

- a photographic reflecting telescope (coronagraph, Huggins system), built by Grubb of Dublin;
- a reflecting telescope (Cassegrain system), constructed by Secretan of Paris;
- a telespectroscope, built by Browning in London;
- a macromicrometer, built by Gautier of Paris;

Fig. 11 The *Carte du Ciel* astrograph in Catania (INAF-OACt Archive)

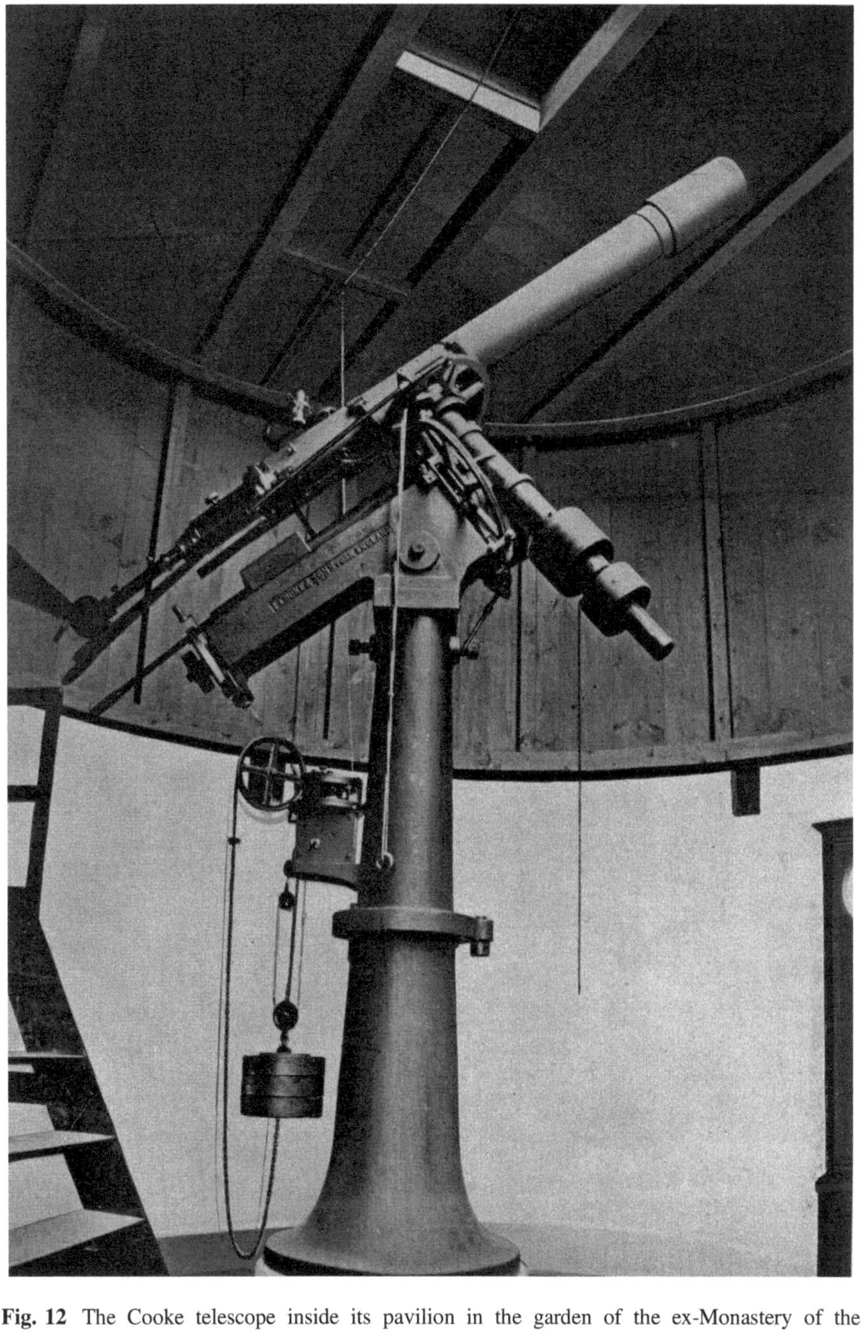

Fig. 12 The Cooke telescope inside its pavilion in the garden of the ex-Monastery of the Benedictines (INAF-OACt Archive)

Fig. 13 The Ertel meridian circle in the garden of the ex-Monastery of the Benedictine (INAF-OACt Archive)

- a diffraction spectroscope, built by the Brassart brothers in Rome;
- a spectrograph (Vogel system), built by Toepfer of Berlin;
- a theodolite, built by Secretan of Paris;
- a steps prism, constructed by Secretan of Paris;
- a prism rangefinder, built by Gautier of Paris;

- a heliostat (Silbermann system), constructed by Secretan of Paris;
- a micrometer scope of Rochon;
- two astronomical pendulums, built by Cavignato of Padua;
- three marine chronometers (Dent, Roskell, Frodsham);
- one dry chronograph (Fuess-Cavignato).

Unfortunately, most of this important scientific instrumentation has been lost, and what remains is still awaiting proper restoration and enjoyment. It is likely that only a contemporary museum space could afford to make known to the general public the history of the Catania Astrophysical Observatory and the legacy left by the astronomers who contributed to its scientific reputation.

Acknowledgements The author thanks Prof. Giovanni Strazzulla (Catania Astrophysical Observatory), Dr. Angela Mangano (Catania Astrophysical Observatory Archive), Dr. Ileana Chinnici (Palermo Astronomical Observatory), Dr. Simona Iozzia (State Archive of Catania) and Dr. Valeria Zanini (Padua Astronomical Observatory) for their kind readiness and valuable collaboration.

References

Abetti G., *Annibale Riccò. 1844–1919*, Astrophysical Journal, vol. 51, pp. 65–72, 1920.

Chinnici I., *La Carte du Ciel. Correspondence inédite conservée dans les Archives de l'Observatoire de Paris*, Observatoire de Paris & Osservatorio Astronomico di Palermo G. S. Vaiana, Palermo, 1999.

Mangano A., *Azeglio Bemporad: un caso di discriminazione razziale all'Osservatorio di Catania*, Giornale di astronomia, Pisa: Fabrizio Serra, pp. 33–36, 2013.

Millosevich E., *L'Osservatorio Bellini sull'Etna*, Roma: Tipografia Barbera, 1880.

Riccò A., *Gli Osservatori di Catania e dell'Etna, Memorie della Società degli Spettroscopisti Italiani*, pp. 144–160, 1898.

Riccò A., *Lo spettroeliografo del R. Osservatorio di Catania*, Rendiconti della R. Accademia dei Lincei, Vol. XVII., serie V, fasc. 12, pp. 529–535, 1908.

Tacchini P., *Eclissi totali di Sole del dicembre 1870, del maggio 1882 e 1883, e dell'agosto 1886 e 1887*, Roma: Tipografia Eredi Botta, 1888.

Tacchini P., *Della convenienza ed utilità di erigere sull'Etna una stazione astronomica-metereologica*, Accademia Gioenia, Catania, 1876.

Double Star Measurement with the Merz Refractors at the Padova and Torino Observatories

Giuseppe Massone

1 Introduction

Double star astronomy effectively began in the first half of the 19th century. After some sparse contributions from some forerunners, the first systematic double star survey, covering the northern hemisphere, was carried out by Friedrich G. Wilhelm Struve (1793–1864) at the Dorpat (now Tartu) Observatory. The results were published in St. Petersburg, in 1837, in the renowned volume *Mensurae Micrometricae* (Fig. 1). This "magnum opus" contains the details of about ten thousand micrometric measurements and an analysis of all available observations of systems showing orbital motion; moreover, the tables of data are preceded by a thorough introduction describing the instrument and observational method used. This introduction became a reference for establishing the standard procedure for double star observation, adopted for over a century by almost all observers, from then on. The refractor telescope was equipped with a filar micrometer designed by Fraunhofer and with Ramsden eyepieces that had spider treads on their focal planes. This was the combination chosen for carrying out visual astronomical measurements throughout the entire 19th century and is still used, with minor modification, nowadays.

The measured coordinates are given by the positional angles counted counter-clockwise from North to East, ranging from 0° to 360°; distances are usually measured in seconds of arc, after the micrometric screw pitch has been calibrated by some angular standard, like the polar or (less frequently) equatorial star's transit. The combination of the observations, when a suitable orbital arc has been observed, yields the apparent orbit of the secondary (fainter) star with respect to the primary.

G. Massone (✉)
INAF-Osservatorio Astrofisico Di Torino, Turin, Italy
e-mail: massone@oato.inaf.it

I. Chinnici (ed.), *Merz Telescopes*, Historical & Cultural Astronomy,
DOI 10.1007/978-3-319-41486-7_9

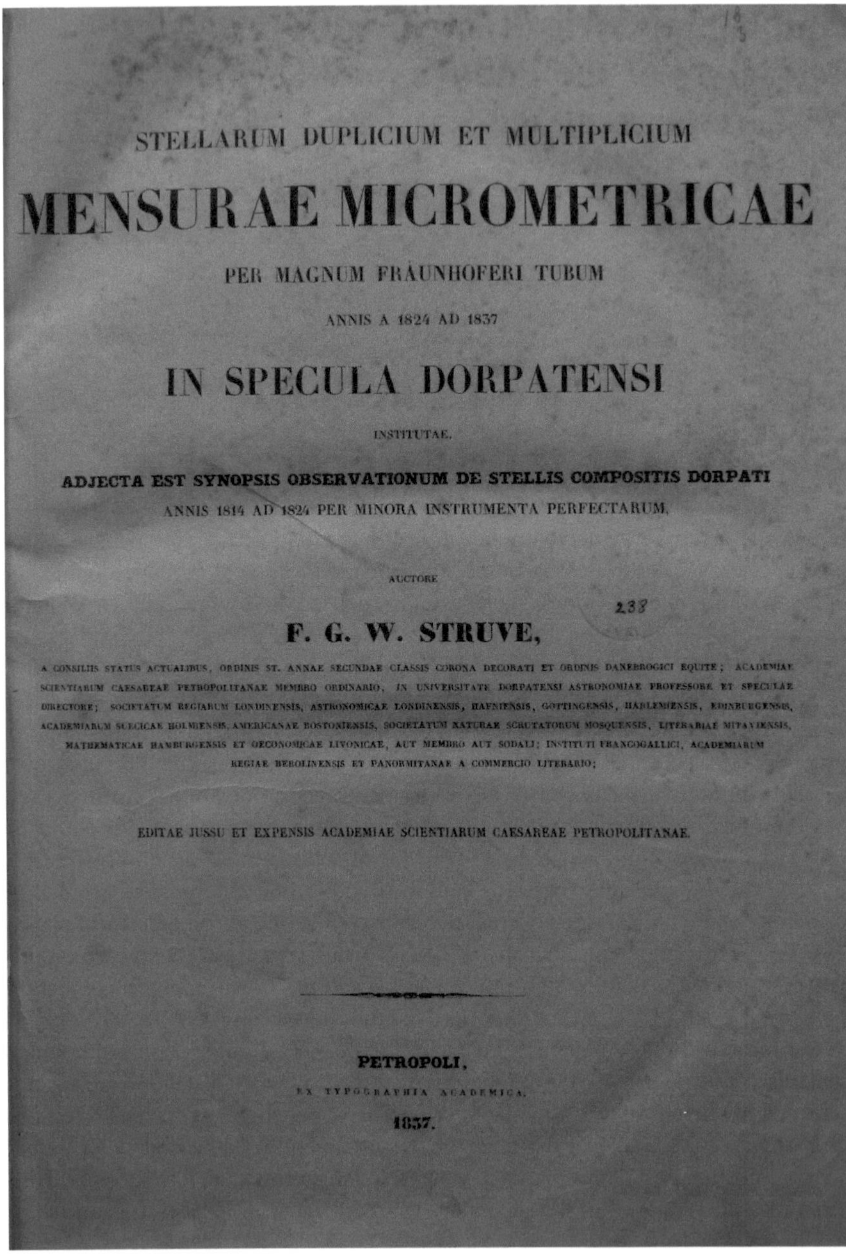
STELLARUM DUPLICIUM ET MULTIPLICIUM

MENSURAE MICROMETRICAE

PER MAGNUM FRAUNHOFERI TUBUM

ANNIS A 1824 AD 1837

IN SPECULA DORPATENSI

INSTITUTAE.

ADJECTA EST SYNOPSIS OBSERVATIONUM DE STELLIS COMPOSITIS DORPATI

ANNIS 1814 AD 1824 PER MINORA INSTRUMENTA PERFECTARUM,

AUCTORE

238

F. G. W. STRUVE,

A CONSILIIS STATUS ACTUALIBUS, ORDINIS ST. ANNAE SECUNDAE CLASSIS CORONA DECORATI ET ORDINIS DANEBROGICI EQUITE; ACADEMIAE SCIENTIARUM CAESAREAE PETROPOLITANAE MEMBRO ORDINARIO, IN UNIVERSITATE DORPATENSI ASTRONOMIAE PROFESSORE ET SPECULAE DIRECTORE; SOCIETATUM REGIARUM LONDINENSIS, ASTRONOMICAE LONDINENSIS, HAFNIENSIS, GOTTINGENSIS, HARLEMIENSIS, EDINBURGENSIS, ACADEMIARUM SUECICAE HOLMIENSIS, AMERICANAE BOSTONIENSIS, SOCIETATUM NATURAE SCRUTATORUM MOSQUENSIS, LITERARIAE MITAVIENSIS, MATHEMATICAE HAMBURGENSIS ET OECONOMICAE LIVONICAE, AUT MEMBRO AUT SODALI; INSTITUTI FRANCOGALLICI, ACADEMIARUM REGIAE BEROLINENSIS ET PANORMITANAE A COMMERCIO LITERARIO;

EDITAE JUSSU ET EXPENSIS ACADEMIAE SCIENTIARUM CAESAREAE PETROPOLITANAE.

PETROPOLI,

EX TYPOGRAPHIA ACADEMICA.

1837.

Fig. 1 The frontpage of Struve's remarkable work on double stars (INAF-OAT Library)

To obtain the true mass of each component, additional information is needed, i.e., parallax and semi-axis ratio, obtained either astrometrically or spectroscopically. However, for the vast majority of double stars, the orbital periods are very long, and reliable orbits have been computed for only a small percentage of known systems. The efforts of several generations of passionate observers has therefore been required, though this kind of work has never gained popularity among astronomers, due to the needful combination of some requisites a good telescope, low atmospheric turbulence for measuring the closest pairs, and the extensive training needed to achieve the ability to properly aim the micrometer treads at the dancing spot of dim light, in the telescope field of view, during the vanishing moments of good atmospheric turbulence. A good historical sketch of that heroic epoch, with quotations from almost all astronomers who spent a significant part of their life to improve our knowledge in this field of research, can be found in Aitken (1935), together with a good introduction to the classical methods of visual observation.

The golden age of the visual observer is now almost completely over: the current techniques, having far greater accuracy and efficiency, are preferred by the new generation of astronomers. However, visual observations of close double stars have been, and still are, in some respect, a good benchmark for assessing the optical quality of a telescope's objective, by exploring, under excellent seeing[1] conditions, the shape and symmetry of diffraction images. The observational data collected by the author of this paper, after some years of visual observation of double stars with two Merz refractors, at the Padua and Turin Observatories, will be used to illustrate once more the outstanding quality reached by this optical firm in the production of astronomical lenses.

2 Telescope Resolving Power and Measurement Technique

The basis of image formation by an optical device, from both the geometric and diffraction points of view, is a fundamental chapter in physics textbooks: a plane wavefront from a point-like source, hitting a circular aperture which borders a converging "perfect" objective, is focused in a "diffraction image" of the source. This is by no means point-like but, though being rotationally symmetrical, it bears a more complex pattern. A central spot, accounting for about 85% in incident light, is surrounded by an infinite number of concentric rings of increasing radius and decreasing surface brightness. The image of an extended object is therefore given by a set of overlapping elementary images.

[1]The word "seeing" is widely recognized among astronomers to indicate the amount of atmospheric turbulence. It is either quantitatively measured in seconds of arcs or qualitatively rated as "poor", "good", "excellent" and so on.

Table 1 The RP for the telescopes used by the author

	d (cm) RP (″)	
Padua Merz telescope	19	0.73
Turin Merz telescope	30	0.46

For a double star with components of similar brightness, the visibility of the two diffraction images from each star of the pair depends on the angular separation of the couple and on the size of the telescope. Regarding the latter, a conventional parameter is used: the Resolving Power (RP), usually defined as the separation angle of two point-like sources when the central spot of the second source falls in the middle of the first dark ring of the first one. The diffraction theory yields the well-known formula:

$$RP = 1.22\frac{\lambda}{d}$$

λ and d being, respectively, the wavelength of the incoming light and the diameter of the telescope's objective.[2] The result is therefore given in natural angle units (radiants). The RP for the telescopes used by the author are indicated in Table 1, assuming a reference wavelength of 0.55 μm, i.e., the one corresponding to the maximum color sensitivity of the human eye.

The actual observation is complicated by several factors: the light is not monochromatic, and therefore the observation is affected by the residual chromatic telescope aberration; the pair's components are of different (sometimes very different) brightness and color; the images are always altered by atmospheric and dome seeing[3] and so on. A more detailed discussion of the resolving power of a telescope, specifically related to visual double star observations, can be found in Couteau (1978), one of the last most experienced observers.

The device used extensively for visual double star observations is the filar micrometer. Other kinds of micrometer have been proposed in different epochs, but none ever gained the popularity of the former. In Padua and Turin, all measurements were done with this instrument, and consequently the technique of its use will be briefly described here. The usual structure of the device is a box placed in the focal plane and rotating around the optical axis of the telescope; it contains three fine wires (usually coming from a spider cocoon): two wires are parallel to each other and perpendicular to a third one; moreover, one of the two is movable by means of a micrometric screw. The degree of rotation of the box is measured on a position circle; the zero setting is achieved by stopping the telescope's motion and aligning the third wire with the sidereal star path: the position angle is obtained by aligning the couple with this same wire. To obtain the couple's distance, a true

[2]The same formula is also used for reflective optics.

[3]The "dome seeing" is the turbulence generated by air temperature gradient between internal and external of the dome.

bisection of the star images, with the micrometric threads, is necessary; this is quite feasible when the two stars are well separated because, for constructional constraint, the diameter of the tread is comparable with that of the central Airy disk of the diffraction image; conversely, this becomes difficult, or even impossible, when the separation is close to the telescope's resolving power, with the Airy disks in contact or partially overlapping. This is quite a frequent occurrence as, statistically, the closer the couple, the faster the orbital motion and, consequently, the higher the astrophysical interest. In these situations, the combination of the observer's skill and the perfection of the telescopic image is critical, as the observer usually overcomes these difficulties in estimating the distance by using the central disk's diameter and the first diffraction ring radius of the primary star as a caliper. Of course, in such a procedure, both systematic and personal errors may arise; this matter was extensively discussed by, i.e., Baize (1949) and Pannunzio et al. (1986, 1988), usually leading to the conclusion that the more experienced the observer, the less likely it is that this kind of error will occur.

3 The Padua Merz Refractor

The Merz refractor of the Padua Observatory had already made a major contribution to double star measurements thanks to the non-professional astronomer Ercole Dembowski (1812–1881).[4] He acquired the refractor in 1860 and devoted himself to an extensive work on double stars: from 1862 to 1878, he used this telescope, and in the years 1851–1858, he used a smaller one, totaling about 21,000 measures. The majority of these data were published posthumously, in two volumes (Dembowski 1883 and 1884) edited by Otto Wilhelm Struve (1819–1905) and Giovanni Virginio Schiaparelli (1835–1910). Dembowski's private observatory was located at Gallarate, in Northern Italy; the Merz telescope was moved to Padua in 1882, once it had been purchased from Dembowski's heirs. There, in a location close to the historical site of the Observatory, where it ended up being used until the 1980s, it became colloquially known as "*The Dembowski*" (Fig. 2), as acknowledgement of the great work on double stars performed by its former owner (see Zanini, in this volume).

The refractor had an aperture of 19 cm and a focal length of 320 cm; it was equipped with the original Merz micrometer and eyepieces. The original mounting, with its beautiful wooden tube and sidereal clock motion, was no longer extant, having been replaced by a new mounting with electric motors. In parallel to the refractor tube, there was a 30-cm-aperture Cassegrain reflector, but this configuration, though a bit more cumbersome than the previous one, was however utilizable for double star work. The new mounting supported the Merz objective-glass, of

---

[4]Dembowsky was a full time astronomer, though without a permanent position in a "professional" observatory.

Fig. 2 The Merz telescope (called *The Dembowski*) at the Padua Observatory in the 1970s (HAOPd)

course, along with the whole old breech block with focusing mechanism and finder telescope. The filar micrometer (Fig. 3) was still in very good condition, only requiring replacement of the broken wires: new ones from a first rate spider cocoon were installed, and the opportunity was taken for a thorough cleaning of the whole device.

The observations carried out with this telescope spanned a period of about two and a half years, with some interruptions, from time to time, due to the student status of the observer. The objects for the observational program were selected by considering the lack of experience of the operator; consequently, the preference was accorded to double stars whose separation was well above the telescope's resolution; however, in some cases, this limit was occasionally approached, generally in accordance with the increasing confidence of the observer.

Fig. 3 The filar micrometer of the *Dembowski* telescope (HAOPd)

4 The Turin Merz Refractor

At the beginning of 1982, the author became employed at the Turin Observatory in a staff position. There, a large refractor was in operation (Fig. 4), with an objective 30 cm in diameter and 445 cm in focal length, acquired from the Merz firm in 1884. The first description of this instrument was published by Alessandro Dorna (1825–1887), director of the Turin Observatory from 1865. He did not have the good fortune to be able to use it for a long time, since he passed away in 1886. At that time, the Observatory was still located at the historical site of Palazzo Madama, right in the middle of the town. The environment, surrounded by increasing light and industrial pollution in the second half of the 19th century, was not optimal for astronomical observations and the telescope saw little use until the Observatory's move to the new site in Pino Torinese was completed in 1912. In 1982 the instrument needed some cleaning and refurbishing, as well as some minor repairs. Two filar micrometers, both by Merz, were available: we chose the older one (Fig. 5), whose design was very similar to the one used in Padua, as I was already fully accustomed to its use. There was also an ample choice of eyepieces, but only two, having magnifications of 580 and 760 respectively, were selected, in order to alter the visual appearance of the star images as little as possible.

In Pino Torinese, the seeing conditions were generally much better than in Padua, and therefore, considering the experience already gained and wishing to push the collected data as far as possible towards interesting astrophysical targets, the observational program included a far greater percentage of close pairs.

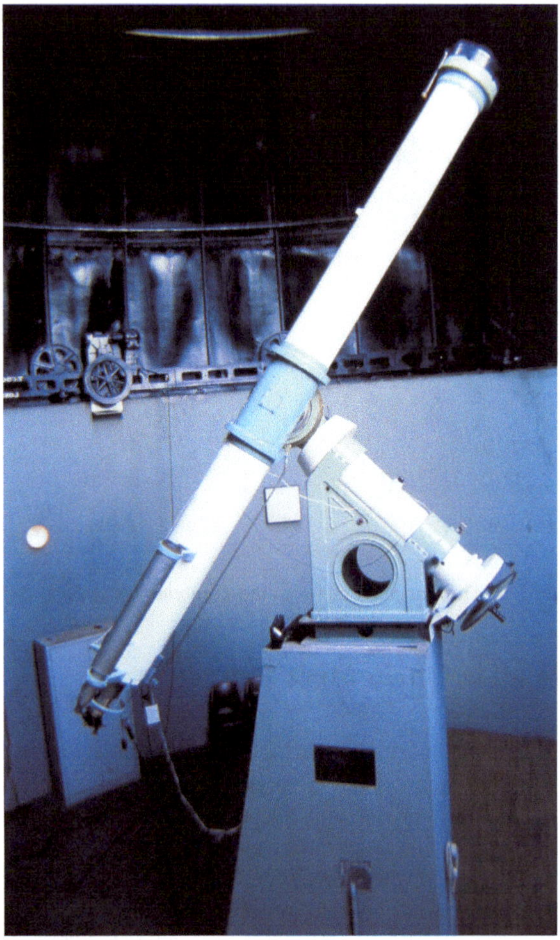

Fig. 4 The Merz telescope at the Turin Observatory in the 1980s (INAF-OAT archives)

The objects selected belonged to three categories: double stars for which a set of orbital elements had already been calculated and for which a regular monitoring of their orbital motion was desirable; double stars that had already shown an appreciable orbital path but for which a set of orbital elements had not yet been computed; and certain objects already observed in Padua but close to the resolving limit of the smaller refractor, therefore requiring observation with a larger instrument.

The observation period began on May 26th, 1982 and was finished by the end of February 1983, because a new and larger refractor had become available, requiring that the Merz be dismantled to free the dome. In this interval, a fairly short one indeed, we collected about 650 micrometric measures; the night details and computed mean for 149 pairs with more than one night of observation available were published (Massone 1985), together with some basic information on the instrument and the observational program.

Fig. 5 The micrometer of the Turin Merz telescope (INAF-OAT collection)

5 Comparison and Evaluation of the Optical Quality of Objectives

Today, the most commonly used method for properly assessing the optical correction and imaging quality of an astronomical objective-glass is the interferometric one; in our case, however, it was not possible to use it, both because of a lack of suitable instrumentation and high costs. Therefore, to compare the performances of the two Merz objectives-lenses, we were only able to rely on the notes taken during our past observational activity with these two telescopes and on the overall data collected. We seldom observed other objects, like the Moon and the planets, but the evaluation that can be inferred from the observations of this class of extended object is much more subjective in comparison with the examination of star images, because, in this case, the differences between apparent and theoretical diffraction images can be evaluated with a good degree of accuracy. We will use two criteria in our tests: the shape and symmetry of diffraction images, as taken from our notes, and a statistic on the distribution of the observed separations.

5.1 Tests on the Dembowski *Telescope*

For this instrument, I do not have detailed notes on the appearance of the diffraction images; however, since it was the first telescope that I was able to use

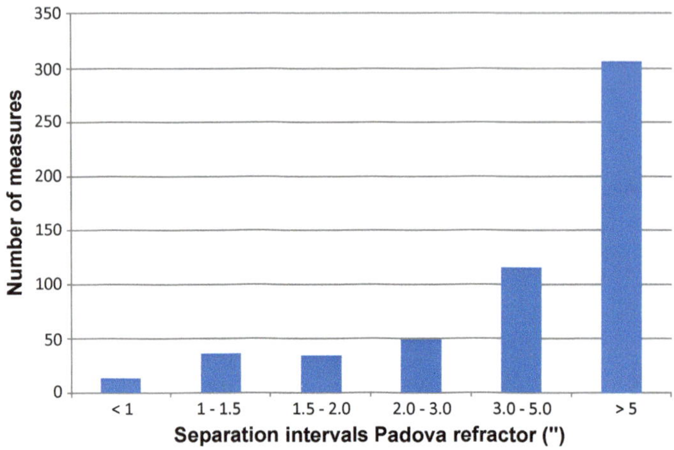

Fig. 6 Statistics of measurements made with the Padua refractor

"professionally", I still maintain a vivid memory of its performance. Though in Padua the seeing conditions were very scarcely optimal for this kind of observation, which requires almost motionless images to appreciate the deviation from the circularity of the central diffraction disk and rings—the latter being most sensitive to residual optical defects—no significant imperfections had been observed, even with the magnification pushed up to 670. The only problem, well known for Merz objective-lenses in general, as indicated, for example, by Schiaparelli (1888 and 1909) for both Merz refractors in Milan, was a chromatic correction mainly weighed for the yellow-red region of the spectrum, rather than balanced over the whole visual range, as is preferred today. However, after a few attempts using illuminated wires for measurements, I switched on the field illumination, as it was most suitable, in my opinion, for ensuring higher accuracy. With this configuration, the faint blue-violet blemish of the secondary spectrum was almost unnoticeable.

The observation of double stars started on June 28th, 1977, and was continued, with some interruptions, until February 19th, 1980, totaling 590 measures. A measure was usually the mean of 4 readings of the position angle and 4 readings of the distance. The distribution of the angular separation for the ensemble of Padua observations indeed, does not do full justice to the optical quality of the Merz refractor: the statistic shown in Fig. 6 is rather the result of the already-mentioned lack of expertise and unexceptional seeing conditions that affected this author's first experience in observing double stars. A much better evaluation can be inferred from Dembowski's original work, in which measures of pairs, with angular separation well below the conventional resolving power of the telescope, were present in abundance, and to which we address the interested reader.

5.2 Tests on the Merz Telescope in Turin

In 1982, the Merz refractor in Turin was not fully operational: the objective-glass had been removed from the tube in order to preserve it (Fig. 7). There were plans to replace the telescope with a new refractor, but the mounting for the latter was not yet available. It was therefore decided that the Merz telescope would be reinstalled in the meanwhile and that a program of double star measurements would be begun, to be continued later with the new refractor. The first observation was carried out on May 17th; here is a quotation from my notes: "… the images are good, though not completely satisfactory. There is the same chromatic shift towards the yellow-red spectral region, already noticed in the Padua 7-inch refractor, but here there is also a high amount of diffuse light; it has not been possible to clean the internal surfaces of the objective lenses, as we do not have the tool to perform this operation safely: we will see how much this may affect the measures of couples with faint companions […] in spite of this circumstance and in spite of the fact that the objective-glass is not yet well aligned, the observations of such double stars as ε Bootis and ζ Bootis show diffraction images without appreciable defects. ζ Bootis components, for instance, were very well separated, and the dark space between the two central diffraction disks was almost of the same length as the diameter of one of the two disks". On the subsequent nights, the optical axis of the objective-lens was finely

Fig. 7 The objective-lens of the Turin Merz telescope (INAF-OAT collection)

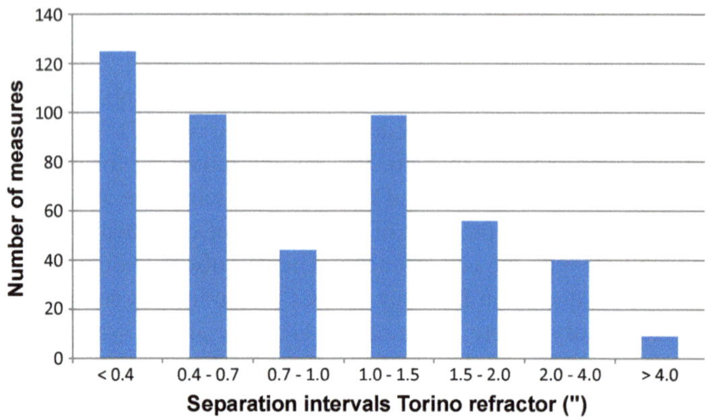

Fig. 8 Statistics of measurements made with the Turin refractor

tuned in order to obtain the best possible images, and a note made on the night of June 2nd is also of interest: "… on the recent nights of May 24th and 25th, the objective-glass has been aligned […] the result, though perhaps not yet perfect, is very satisfactory, because the diffraction image does not show any asymmetry, having a well-rounded central disk."

The distribution of the observed angular separations measured at Pino Torinese show that the performance of the Merz objective-lens approached the real observing limit of the instrument, much more so than in the case of Padua. In Pino Torinese, the seeing conditions were really good for an appreciable number of nights, and the observer, perhaps with a bit of overconfidence, considered himself already weaned from this kind of work. The histogram in Fig. 8 shows this clearly, with a large number of pairs having a separation well below the resolving power of the telescope. On nights with low atmospheric turbulence, the lack of roundness on the central Airy disk for a very closed double was still perceivable for separations down to 0″.25–0″.26.

References

Aitken, R. G.: 1935, "The Binary Stars", McGraw-Hill Book Company, Inc.

Baize, P.: 1949, Jour. des Obs., vol. XXXII, n. 8–9, pag. 97

Couteau, P.: 1978, "L'Observation des Etoiles Doubles Visuelles", Flammarion, Paris

Dembowski, E.: 1883, "Misure micrometriche di stelle doppie e multiple", Atti della Regia Accademia dei Lincei, anno CCLXXX, Memorie della Classe di Scienze Fisiche, Matematiche e Naturali, vo.l XVI, Roma

Dembowski, E.: 1884, "Misure micrometriche di stelle doppie e multiple", Atti della Regia Accademia dei Lincei, anno CCLXXXI, Memorie della Classe di Scienze Fisiche, Matematiche e Naturali, vol. XVII, Roma

Massone, G.: 1985 Astron. Astroph. Suppl. Ser., Vol. 60, pag. 485

Pannunzio, R., Zappala, V., Massone, G., Morbidelli, R.: 1986, Astron. Astroph. Vol. 166, pag. 337

Pannunzio, R., Massone, G., Morbidelli, R.: 1988, Astron. Astroph. Vol. 203, pag. 388

Schiaparelli, G. V.: 1888, "Osservazioni sulle Stelle Doppie", serie prima comprendente le misure di 465 sistemi eseguite col refrattore di 8 pollici di Merz negli anni 1875–1885, Pubbl. del Reale Oss. di Brera in Milano, N. XXXII

Schiaparelli, G. V.: 1909, "Osservazioni sulle Stelle Doppie", serie seconda comprendente le misure di 636 sistemi eseguite col refrattore equatoriale Merz-Repsold negli anni 1886–1900, Pubbl. del Reale Oss. di Brera in Milano, N. XLVI

Struve, F. G. W.: 1837, "Stellarum Duplicium et Multiplicium Mensurae Micrometricae per Magnum Fraunhoferi Tubum annis a 1824 ad 1837 in Specula Dorpatensi", St. Petersburg

Appendix A
A List of Extant Large Merz Telescopes (aperture => 150 mm) Worldwide[1]

By Paolo Brenni and Ileana Chinnici

Observatory	Size (aperture) (mm)	Signature	Date	Comments
Cambridge	158		1860	Now at Bayfordbury Observatory (Hatfield)
Bogenhausen (Munchen)	267		1835	Initiated by Fraunhofer Now in Universitätssternwarte (Bogenhausen)
Bordeaux	380	Merz-Eichens/Gautier	1883	
Cape Town (South Africa)	178	Merz-Troughton&Simms	1849	Combined with a 18-inch reflector, out of service
Catania	320	G. & S. Merz	1876–77	In restoration
Cincinnati (Ohio, USA)	280	Merz-Mahler	1843	
Copenhagen	300			Now at The Steno Museum
Greenwich	330	Merz	1859	Now at The Observatory Science Centre (Herstmonceux), combined
Hamburg (moved to Begedorf at the	260	Merz-Repsold		Modified

(continued)

[1]*Some sources report about a 270-mm refractor in Madrid (see: https://arxiv.org/ftp/arxiv/papers/1407/1407.3104.pdf, access 4 November 2016) and a 180-mm refractor at Haynald Observatory in Kalocsa, Hungary (see: http://fenyi.solarobs.csfk.mta.hu/Kalocsa/Kalocsa_E.html, access 4 November 2016), but it is unclear if these telescopes are still extant. Many Merz telescopes, whose destiny is obscure, are mentioned in lists compiled at the end of 19th century (i. e., see: Payne, W.W. 1884, "Large Telescopes of the World", *Sidereal Messenger*, vol. 3, pp.193–198)

I. Chinnici (ed.), *Merz Telescopes*, Historical & Cultural Astronomy,
DOI 10.1007/978-3-319-41486-7

(continued)

Observatory	Size (aperture) (mm)	Signature	Date	Comments
beginning of 20th century)				
Harvard (Cambridge, Mass.)	380		1847	
Kiev (Russia, today Ukraine)	245	Merz-Mahler		
Kis-Kartal (Hungary)	190			
Kremsmünster	150	Merz-Starke	1857	
Lisbon	382	Merz-Repsold	1860s	Modified
Lund (Sweden)	240		1867	
Marseilles	255	Merz-Eichens	1872	
Milan (Brera)	218 490	Merz-Repsold	1862–65 1881–82	Restored Restored (lens broken)
Neuchatel (Switzerland)	162			Complete but stored and not mounted
Nikolaev (Russia, today Ukraine)	243	Merz-Repsold		
Padua	187		1862	Restored
Palermo	250	G. Merz und Sohn	1859	Restored, modified
Quito (Ecuador)	229		1875	
Strasburg	162 487	Merz-Repsold	1876 1877–80	
Turin	330		1884–85	Modified, only lens extant
Wildhaven (Boone County)	190	Merz and Mahler	1845	Restored

Appendix B
List of Astronomical Instruments by Merz (1826–1932)[2]

Astro-Objectives

- Baltimore, Maryland (USA), Private Observatory Wright
- Berlin, Royal Observatory
- Bonn, Royal Observatory
- Bordeaux—Floirac (France), University Observatory
- Bruxelles (Belgium), Observatory of the Royal Academy
- Bucarest (Romania), Observatorium for Astrometry
- Catania, Sicily (Italy), Royal Observatory
- Düsseldorf—Bilk, Observatory Charlottenruhe
- Elchies—Morayshire (Scotland), Private Observatory Grant
- Enge (Switzerland), Private Observatory Kann
- Fairford (England), Private Observatory Bazley
- Geneva (Switzerland), University Observatory
- Gent (Belgien), Private Observatory Monckhoven
- Greenwich (England), Royal Observatorium
- Hamburg, Observatory
- Hamburg—St. Pauli, Straßenastronom Lindow
- Ipswich (England), Orwell Park Observatory
- Kis-Kartal (Hungary), Private Observatory Podmaniczky
- Kopenhagen (Danmark), University Observatory
- Landstuhl, Private Observatory Fauth
- Lima (Peru), Staatl. Observatory
- Lisbon (Portugal), Royal Observatory
- Lund (Sweden), University Observatory
- London (England), Private Observatory Pritchard
- Milan (Italy), Royal Observatory Brera
- Manila (The Philippines), Observatory of the Jesuit College

[2]By Jürgen Kost Excerpt from: Jürgen Kost, *Wissenschaftlicher Instrumentenbau der Firma Merz in München 1838–1932*, Nuncius Hamburgensis Beiträge zur Geschichte der Naturwissenschaften Band 40, Hamburg: tredition 2015.

I. Chinnici (ed.), *Merz Telescopes*, Historical & Cultural Astronomy,
DOI 10.1007/978-3-319-41486-7

- Marseille (France), University Observatory
- Nikolajew (Mykolajiw, Ukraine), Marine Observatory
- Odessa (Ukraine), University Observatory
- Pulkowa (Pulkovo, Russia), Nikolai Observatory
- Revard (France), Private Observatory Jarry-Desloges
- Rome (Italy), Private Observatory Ferrari
- Rome (Vatican), Vatican Observatory
- Santiago (Chile), State Observatory
- Stockholm (Sweden), University Observatory
- Strasbourg/Straßburg (France), University Observatory
- Turin (Italy), University Observatory
- Utrecht (The Netherlands), University Observatory

Refracting Telescopes and Tubes

- Altona (Denmark, 1867 Prussia, 1938 Hamburg), Royal Observatory
- Berlin, Royal Observatory
- Bern (Switzerland), University Observatory
- Biberach—Rot an der Rot, Private Observatory
- Calcutta, West Bengal (India), University Observatory
- Cambridge, Massachussets (USA), Harvard University
- Camden Lodge, Kent (England), Private Observatory Dawes
- Capetown (South Africa), Royal Observatory
- Cincinnati, Ohio (USA), Cincinnati Observatorium
- Düsseldorf-Bilk, Private Observatory Charlottenruhe
- Frederiksværk (Denmark), Private Observatory Fock
- Fredricton (Canada), University Observatory
- Frome, Somerset (England), Private Observatory Brodie
- Gallarate (Italy), Private Observatory Dembowski
- Göttingen, University Observatory
- Helsinki (Finland), University Observatory
- Kalocsa (Hungary), Jesuit Observatory
- Kasan (Russia), University Observatory
- Kiew (Ukraine), University Observatory
- Copenhagen, for Jünger (Denmark)
- Copenhagen, (Denmark) Observatory "Rundetårn"
- Krakow (Poland), University Observatory
- Kremsmünster (Austria), Benediktiner Observatory
- Leiden (The Netherlands), University Observatory
- London (England), an Jackson
- London (England), Private Observatory Lettsom
- Madrid (Spain), Royal Observatory
- Milan (Italy), Brera Observatorium
- Manila (The Philippines), Observatory of the Jesuitenkollegs
- Moncalieri, Turin (Italy), Meteorological Observatory
- Munich-Bogenhausen, Royal Observatory

- Munich, Private Observatory Strähuber
- Moscow (Russia), University Observatory
- Naples (Italy), Royal University Observatory
- Neuchâtel (Switzerland), Kantonssternwarte
- New Hanover, North Carolina (USA), School Observatory
- Nuremberg/Nürnberg, Private Observatory Eichhorn
- Ofen (today Budapest, Hungary), Royal Observatory
- O'Gyalla (Hungary), Private Observatory Konkoly (today: Hurbanovo, Slowakia)
- Oslo/Christiania (Norway), University Observatory
- Oxford (USA), Private Observatory Harper
- Palermo, Sicily (Italy), Royal Observatory
- Palsgaard (Denmark), Private Observatory Reedtz
- Pardubice (Czech Republic), Public Observatory
- Philadelphia, Pennsylvania (USA), School Observatory
- Philadelphia, Pennsylvania (USA), Private Observatory Justice
- Pulkovo (Russia), Nikolai-Hauptsternwarte
- Pullach, Private Observatory Anton Staus (1872–1955)
- Pulsnitz, Private Observatory Classen
- Quito (Ecuador), State Observatory
- Remscheid-Hasten, for Cleff
- Ribinsk (Russia), Private Observatory Koschin
- Rome (Italy), Capitol Observatory
- Rome (Vatican), Collegio Romano Observatory
- Shelbyville, Kentucky (USA), College Observatory
- Solingen-Merscheid, Private Observatory Linder
- St. Petersburg (Russia), Private Observatory Basilewsky
- St. Petersburg (Russia), University Observatory
- Sydney (Australia), Royal Observatory
- Tokyo (Japan), University Observatory
- Tunbridge Wells (England), Private Observatory McClean
- Tübingen, University Observatory
- Utrecht (The Netherlands), University Observatory
- Vejle (Denmark), Private Observatory Nicolaisen
- Vienna (Austria), University Observatory
- Vilnius (Lithuania), University Observatory
- Warsaw (Poland), University Observatory
- Washington (USA), Marineobservatorium
- Zurich (Switzerland), Polytechnical Observatory

Heliometers

- Bamberg, Dr. Carl-Remeis-Observatory
- Bonn, Royal Observatory
- Cape Town (South Africa) Royal Observatory
- Kaliningrad (Russia), Royal Observatory

- Oxford (England), University Observatory
- Pulkovo, St. Petersburg (Russia), Nikolai Observatory
- New Haven, Connecticut (USA), Yale College

Medial Telescopes

- Landstuhl (Pfalz), Private Observatory Fauth
- Heidelberg, Landessternwarte

Comet Seekers

- Bamberg, Remeis-Observatory
- Dresden, Private Observatory Engelhardt
- Göttingen, University Observatory
- Moscow (Russia), University Observatory
- Strasburg (France), University Observatory
- Vienna (Austria), University Observatory.

Index of Names

I. Chinnici (ed.), *Merz Telescopes*, Historical & Cultural Astronomy,
DOI 10.1007/978-3-319-41486-7

Index of Places

I. Chinnici (ed.), *Merz Telescopes*, Historical & Cultural Astronomy,
DOI 10.1007/978-3-319-41486-7

Index of Celestial Objects

I. Chinnici (ed.), *Merz Telescopes*, Historical & Cultural Astronomy,
DOI 10.1007/978-3-319-41486-7

MIX
Papier aus verantwortungsvollen Quellen
Paper from responsible sources
FSC® C105338

If you have any concerns about our products,
you can contact us on
ProductSafety@springernature.com

In case Publisher is established outside the EU,
the EU authorized representative is:
Springer Nature Customer Service Center GmbH
Europaplatz 3, 69115 Heidelberg, Germany

Printed by Libri Plureos GmbH
in Hamburg, Germany